ADMINISTERING THE COMPANY WAREHOUSE AND INVENTORY FUNCTION

ADMINISTERING THE COMPANY WAREHOUSE AND INVENTORY FUNCTION

Jay W. Spechler, P. E.

PRENTICE-HALL, INC. Englewood Cliffs, New Jersey

Prentice-Hall International, Inc., *London*
Prentice-Hall of Australia, Pty. Ltd., *Sydney*
Prentice-Hall of Canada, Ltd., *Toronto*
Prentice-Hall of India Private Ltd., *New Delhi*
Prentice-Hall of Japan, Inc., *Tokyo*

Library of Congress Cataloging in Publication Data

Spechler, Jay W
Administering the company warehouse and inventory function.

1. Physical distribution of goods--Management.
2. Materials handling. 3. Inventory control.
4. Warehouses. I. Title.
HF5415.7.S68 658.7'8 74-23156
ISBN 0-13-004861-5

Printed in the United States of America

DEDICATION

To my wife, Marilyn, whose support, encouragement and sacrifices have made it possible for it all to come together.

And to my parents who taught me the value of education and hard work.

About the Author

Jay W. Spechler, P.E. is President, Consulting Management Engineers and a Director of the Management Development Institute. He has worked as a Vice President of Manufacturing for a major textile manufacturer and served as an internal consultant for Gulf and Western Industries. Prior to that assignment, he was Director-Materials Management for McCalls Printing Company and Manufacturing and Distribution Operations Coordinator for Century Paper Company.

Mr. Spechler has conducted Manufacturing and Physical Distribution Management seminars for the American Management Association and for Advanced Management Research, Inc. He has also lectured privately and at academic institutions over the years.

His articles on various business management subjects and innovative equipment and systems designs have appeared in numerous publications, and he is listed in *World Who's Who in Commerce and Industry.*

Mr. Spechler holds a B.S. in Production and Management from New York University, an M.S. in Industrial Engineering—Stevens Institute of Technology—and has done extensive post-graduate work in Labor Economics at the University of Houston.

He is a registered Professional Engineer and a member of the American Institute of Industrial Engineers, the National Council of Physical Distribution Management, the International Materials Management Society and the American Economics Association.

What This Book Can Do for You

This book has been written in "how to do it" fashion. It is for *anyone* who has responsibility for reducing and controlling physical distribution and inventory carrying costs.

Great emphasis has been placed on "building block" approaches in achieving cost reduction/control objectives. Drastic, sweeping system changes rarely get results. Rather, objectives are achieved through the systematic approaches in analysis and implementation presented in the following chapters.

Nor is there presented or intended anywhere in the book, "the one best way." Emphasis is placed on illustrating those scientific management techniques which can help you find "the best way" *for your operation.* Numerous examples of analytical technique followed by specific practical illustrations of methods and systems implementations are presented for your consideration.

In short, each chapter aids you in achieving significant cost reduction/control results in your company's individual environment.

GETTING THE EDGE OVER COMPETITION THROUGH EFFECTIVE CUSTOMER SERVICE

This chapter "comes to grips" with the many important aspects of customer service. It presents the most complete set of customer service definitions and measurement techniques to be found anywhere. For example, there is a comprehensive presentation of *how to* measure and reduce customer service cycle time; several different ways to define and calculate customer service levels in percentage terms is also shown.

The latter enables you to give your top management several options in determining the most meaningful yardstick of customer service performance.

Also important to management is how the company's customers perceive the service given to them. A complete customer service survey outline is presented along with a method by which the customer's responses can be translated into dollar savings for you.

Corporate policies which can improve customer service without increasing costs are discussed. Consideration is given to such factors as: balancing customer service levels against individual item contribution to sales; establishing and monitoring lead-time considerations in resupply; order entry procedures; as well as several others.

ORGANIZING THE DISTRIBUTION FUNCTION FOR TOP PROFITS

Alternative Physical Distribution structures are presented which can aid management in organizing P. D.-related functions in a profit maximizing manner. Respective departmental responsibilities are also outlined in detail.

Several pitfalls along the road to establishing a well-organized P. D. department are discussed, along with some suggested remedies.

Specific examples of dollar savings in reduction of machine down-time and in job consolidations, resulting from P. D. organizational improvements, are outlined.

The recently evolved, "Materials Control Total Cost Organizational Concept" is discussed in terms of its profit improvement potential to the individual firm.

COORDINATING MARKET AND FINANCIAL INTERESTS WITH PHYSICAL DISTRIBUTION CONCERNS

Direct and indirect marketing pressures on the P. D. organization and profitability of the firm are described throughout the first half of this chapter. Specific recommendations are presented whereby Physical Distribution can lend significant support to Marketing while not adversely affecting profitability.

Financial/P. D. interaction and dependence is treated in detail. Several ways of treating distribution costs are shown. A suggested list of computer-generated basic management reports for distribution firms is also presented.

Advanced P. D. profitability analysis techniques are described in detail, followed by examples of how companies have successfully applied them.

ELIMINATE PROFIT LEAKS BY IMPROVING YOUR PLANT SECURITY

A systematic approach towards analyzing and implementing improved warehouse security procedures is presented. Good physical security is shown as being dependent upon proper administrative and procedural controls.

An outline with detailed procedures for conducting an internal and external facility security survey of any company is shown. Attention is given to such factors as high-value item security methods, reducing delivery driver shortages, guard force procedures, among many others.

Several case histories of employee theft and other malpractices are presented along with a description of corrective actions.

EXAMINE WAREHOUSE OPERATIONS FOR COST REDUCTION POSSIBILITIES

Cost/benefit improvement emphasis is given to four important operational areas within a Physical Distribution facility:

- How to conduct a systematic survey of existing operations with improvements in methods and control.
- A freight expense analysis system designed to significantly reduce transportation costs.
- Procedural flow chart analysis and other ways to reduce customer order cycle time.
- Examples of proven warehouse control reports used to reduce direct labor expense.

SELECTING MATERIALS HANDLING EQUIPMENT TO FIT YOUR PARTICULAR NEEDS

This chapter simplifies the increasingly complex problem of proper materials handling equipment selection. A unique "Matrix Analysis" chart is presented which will aid the P. D. manager in narrowing his choice to the few best alternatives. Every major category of materials handling equipment is included.

A total-cost approach is taken, with consideration given to first cost as well as continuing operating expenses.

Several examples follow the use of the "Matrix Analysis" in a step-by-step manner.

EFFECTIVE WAREHOUSE LAYOUT PLANNING REDUCES MATERIALS HANDLING COSTS

The material in this chapter assists the P. D. manager in operating at minimum cost while maximizing customer service.

Layout planning features include, in part, attention to the following factors:

- Mechanical aids in material movement.
- Storage equipment to maximize cube utilization.
- Aisle width considerations.
- Pitfalls and benefits in consolidating multiple warehouse locations.
- Stock rotation.

A newly developed "Storage Profile Analysis" procedure is shown which greatly simplifies the calculations of storage space and equipment requirements.

Special attention is given to layout management problems in the small warehouse.

MAKING VALUE ENGINEERING WORK FOR YOU IN REDUCING DISTRIBUTION COSTS

Here, and as far as is known, for the first time, value engineering/control is developed for specific application to Physical Distribution operations.

Instruction is given on how to introduce a "value control" program into your distribution operation. Detailed coverage is given to the following:

- How to present the "value" and profit contribution which P. D. can make to your company.
- A systematic, step-by-step procedure to use in starting a value control program.
- Examples which show how P. D. managers in different industries made "value control" work for them in achieving significant dollar savings.

EFFECTIVELY USING AND MAINTAINING LIFT TRUCKS

Discussed in detail, is a five-part, systematic approach to getting the greatest dollar return on your fork-truck investment:

- An operator program designed for safe and efficient fork-truck use.
- Equipment maintenance.
- Popular attachments which make fork trucks more efficient.
- Selection of pallet racks to match fork trucks.
- A formula to determine when materials handling equipment should be traded.

SURVEYING, FACT GATHERING, ANALYZING, RECOMMENDING AND IMPLEMENTING INVENTORY MANAGEMENT IMPROVEMENTS

A detailed inventory management survey is outlined with the recommendation that it be used to analyze the company's inventory position, the potential for improvement and the firm's capacity to implement changes.

The material is sufficiently comprehensive to permit the P. D. manager or staff personnel to perform the survey without outside assistance. The procedure utilizes tables, graphs, procedural flow charts, computer print-out formats and a PERT chart survey implementation program.

The probability of inventory management system success is improved by following certain rules elaborated upon in this chapter. Particular emphasis is placed on keeping on target through the use of information feedback systems and through a suggested, disciplined approach in processing input data.

PLANNING FOR SUCCESSFUL RESULTS IN INVENTORY MANAGEMENT

A very comprehensive discussion covering all forms of inventory control and inventory management systems is presented in the first half of this chapter. The complete spectrum of inventory systems from manual through the most sophisticated computerized version is described, with advantages and disadvantages set forth. Graphs are employed to clarify vital points, and all formulas are accompanied by actual examples to eliminate any doubt as to how they are used in the real business world.

The second half of the chapter is devoted to sales forecasting—the most important planning factor in inventory management. All of the practical methods of forecasting are presented. Both manual and computer approaches are described in detail. Finally, actual examples of benefits which companies have obtained through the use of these forecasting techniques are shown.

COUPLING GOOD INVENTORY TURNOVER WITH BETTER CUSTOMER SERVICE

This chapter shows you how to develop and implement a total inventory management reporting system. These reports can provide you with total control over your inventory performance profile. Included in this section are:

- How to define and calculate inventory turnover.
- How to calculate months of supply.
- How to evaluate inventory turnover vis-a-vis customer service.
- How to measure your turnover against the rest of your industry.
- How to prepare a customer service level feedback report.
- How to prepare an inventory aging report.
- How to prepare a distribution by value report.

AVOIDING PITFALLS IN INVENTORY SYSTEMS

Knowing why and how others failed in their efforts to establish inventory management systems is viewed as a vital ingredient in assuring your own success. Pains are taken to provide specific advice as to how to overcome obstacles in the path of successful inventory management system implementation. Some of the major points covered are:

- Faulty system design.
- How to best report sales history.
- Establishing accurate lead-times.
- Exceptions reporting.
- Output report frequency.

In addition, many other vital considerations are posed.

SELECTING A WAREHOUSE LOCATION HAVING THE BEST PROFIT POTENTIAL

This chapter shows you how to:

- Prepare a manual economic analysis of site selection alternatives which shows you the least costly site.

- Calculate costs where you intend to build your own facility rather than lease.
- Determine when to use computer-assisted analysis in selecting a site.
- Develop computer formats and print-outs used in site selection.

Jay Warren Spechler

ACKNOWLEDGMENTS

Those executives whose names appear below contributed to making this book a reality. Their objectivity, their confidence in my work and concepts and their willingness to risk their own reputations on innovative approaches were important ingredients in my development.

These men think of themselves as businessmen and they are the best of that breed. But they are also true educators who impart their practical and honest business philosophies to their associates.

John Anthony	Director The Personnel Counselors Houston, Texas
Herschel Maltz	President Century Paper Company Houston, Texas
Jules Maltz	Chairman of the Board Century Paper Company Houston, Texas
Melvin Maltz	Vice President Century Paper Company Houston, Texas
Paul Miller	Vice President U.S. Filter Corporation New York, New York
Robert C. Phelan	Senior Vice President Consolidated Cigar Corporation New York, New York
Elio Rotolo	Director ITT New York, New York

Al Schneider	Principal Schoolman Consultants, Inc. Setauket, New York
William Schoolman	President Schoolman Consultants, Inc. Setauket, New York

J. W. S.

Table of Contents

1

Getting the Edge Over Competition Through Effective Customer Service

A SYSTEMATIC, PROFIT BOOSTING APPROACH TO MEASURING YOUR COMPANY'S CUSTOMER SERVICE PERFORMANCE

Achieving a high level of customer service ranks first on the list of important physical distribution objectives.

If your company has competitors, then your sales force is selling customer service right along with the product itself. In fact, customer service has become a dominant factor in the majority of Purchasing departments. This is due to the fact that more and more companies are demanding that their suppliers carry a larger proportion of their inventory requirements for them. This reduces the purchaser's inventory investment while at the same time requiring that the supplier carry more stock and fill orders more frequently.

As vital a consideration as customer service is, few firms have a means of continuously measuring it; have defined what aspects of customer service they can control; have systematically evaluated the potential for customer service position compared to that of their competitors. This chapter gives you the practical techniques for measuring and controlling customer service, enabling you to apply these methods in your situation.

Since any policy changes on customer service will have a significant impact on distribution costs, it is most important that customer service objectives be stated within the context of what your customers view as being a reasonable level of service.

CASE IN POINT: Before having customer service performance measurement tools available to it, a major food products distributor reacted to individual customer service complaints by substantially increasing inventories. After installing a back-order measuring system similar to that shown in Exhibit 1-1 (page 32), the company found that it was shipping 99% of items ordered within 24 hours. This is, of course, an excellent record, but the competition was providing only 95% service and the company's own customers said (when surveyed) that they would be satisfied with 95% item fulfillment. Going from a 99% to a 95% service level enabled the firm to reduce its inventory level by $1,200,000, without losing sales. The key to this company's success in accomplishing this inventory reduction was educating and working closely with its customers, and then making its new customer service measurement tools work for it.

The key hallmarks of superior customer service are:

(1) *Short lead time*—the elapsed time from order inception by a customer to his receipt of goods.

(2) *Consistency of performance*—within published lead times.

(3) *Availability of order information*—promptly, accurately, courteously.

Customer service, therefore, is affected by sales policies, information policies, information processing, the geographic orientation of the distribution network, the physical movement of goods and the availability of goods.

The qualitative aspects of customer service, such as billing practices and communications accuracy, are usually well scrutinized. This chapter is, therefore, largely devoted to developing specific quantitative working tools which will enable you to improve, control and evaluate your company's customer service performance.

The quantitative aspects of customer service may be divided into the following three parts:

1. *Service Cycle.* The time from placement of orders to the moment that merchandise is delivered to your customer's location.
2. *Service Level.* The required degree of reliability in achieving this level.
3. *Order Completeness.* The percentage of orders shipped complete.

A. CUSTOMER SERVICE SURVEY

A survey of customers' perceptions of a company's order fulfillment performance provides vital intelligence to the Physical Distribution and Marketing departments. It enables the company to objectively evaluate its own customer service performance against competition in different geographic areas and is essential in establishing customer service level objectives, inventory levels and the number and location of distribution centers. In addition, the Marketing department usually considers such surveys as an opportunity to improve customer relations. It also often proves useful in identifying significant areas within the distribution operation which need improvement. For example, one survey revealed that customers were not obtaining satisfactory answers to customer service inquiries. They complained that order status and inventory availability information was not consistently and accurately available to them. The company took corrective action. Better order status and production planning information reports were designed, and a special group of knowledgeable people were assigned to handle customer inquiries.

For best results, the survey should be conducted by an outside consulting firm. This serves to eliminate any bias which your own sales force or other personnel might introduce into the questions and/or responses. Further, customers tend to answer sensitive questions more openly and provide more data when they know that confidentiality will be maintained.

An outline of a Customer Service Survey, showing the important subjects to be covered and detailed questions to be answered, is shown below. It may be easily tailored to fit any company's requirements.

*Interview Number*____________

CUSTOMER SERVICE SURVEY

Name of distributor (customer) ____________________

Address ______________________________

City __________________ State ____________

Respondent ______________ Title ____________

Account volume $ ____________

District __________________

Evaluation

Excellent ______

Good ______

Fair ______

Poor ______

Cooperation

Excellent ______

Good ______

Fair ______

Poor ______

Company brands carried ____________________

____

____

Other brands carried ____________________

____

____

Rank your suppliers in order of importance to your operation.

1. ____________ ____________
2. ____________ ____________
3. ____________ ____________
4. ____________ ____________

I. INVENTORY POLICY

A. Describe your basic inventory policy for the company's products. How many turns do you aim for?
B. How does this differ for fast and slow moving items?
C. Is it different for items other than those supplied by our company?
D. Does your inventory policy vary by supplier? If yes, how does it vary?
E. Do you place your orders according to a set schedule? If yes, what is it?

II. ORDER PLACEMENT PROCEDURES

A. How are regular orders initiated?
B. Who does the actual order placement?
C. How are the orders transmitted?
D. (If by mail): How long do you allow for the mails?
E. Do your order placement procedures vary when you order special merchandise or promotional items?
F. Are these procedures any different when you order from other suppliers?
G. Do you ever change orders after placement?
H. Do you do this often?
I. How does the company respond to order changes?
J. Does the company's response differ from your other suppliers'?

III. CYCLE TIME

A. What is your average cycle time? (Order placement to receipt of merchandise.)
B. Do you keep records of this? If yes, what type of records?
C. Is the company's order cycle reliable? If not, how wide are the swings?
D. What is the order cycle time for your other suppliers? Are they reliable?

Supplier	Cycle Time	Reliability
__________	__________	__________
__________	__________	__________
__________	__________	__________
__________	__________	__________

E. Does the ease and speed with which you can place orders have any significant effect on your purchase decisions?
F. Should the company speed up its order cycle time?
G. If yes, what would you like it to be?
H. What effect would this have on your inventories?
I. How would this effect the efficiency of your operation?
J. Would this have any meaningful effect on your profit performance?
K. Would it affect your total purchases of the company's products?

IV. SHIPPING

A. Does the company ever make shipping mistakes?
B. (If yes): What kind of mistakes are most frequent?
C. How often do these occur?
D. Do you keep records on shipment accuracy?
E. How does the company compare with its competition in this regard?

V. INQUIRIES

A. Do you ever inquire about the status of an order with the company?
B. What are the usual reasons for the inquiry?
C. How often do you make inquiries?
D. Do you get the information you need quickly?
E. Is it accurate?
F. How do you inquire about order status with other suppliers?
G. How does the company compare with competition in terms of speed and accuracy of order information?

VI. ORDER COMPLETENESS

A. Does the company normally ship you complete orders?
B. Do you monitor supplier performance in this regard?
C. How often do they back-order?
D. Does this amount of back-ordering affect the company's overall volume with you?
E. How long do you wait for back-ordered shipments? Do you keep records?
F. How are back-orders processed by your system?
G. How does the company compare with its competition on frequency of back-order, on the length of time it takes them to fill back-orders?
H. If the company shipped 100% complete orders, would your purchases of their product increase significantly?

VII. DELIVERY

A. Do you ever have problems with the company's deliveries?
B. Are they made at convenient and predictable hours?
C. Are carrier personnel helpful and courteous?
D. Do you have preferred routings for your shipments? Does the company follow your routing requests?
E. Are the company's shipments ever damaged en route? (If yes): How often does this occur?
F. Is spoilage a problem with the company's products?
G. Do you think the company packages its products well?
H. Do you ever have difficulties in receiving the company's merchandise?
I. Could you take in larger orders than you do now?

VIII. COMMUNICATIONS

A. Who do you contact at the company when you need to make an inquiry, correct a mistake or register a complaint?
B. Are the company's customer service personnel courteous and helpful?
C. Do they have access to needed facts? How long does it take to receive requested information?
D. How do you communicate with your other suppliers?
E. How does the company compare with its competition in this area?

IX. BILLING

A. Do you ever have problems with the company's billing and invoicing procedures?

B. Does the company present its bills promptly?
C. Does the company ever make billing mistakes?
D. (If yes): Do they occur often enough to be a serious problem?
E. Are adjustments and credits handled quickly and accurately?
F. How long does it take to correct a mistake and clear the necessary paperwork?
G. How does the company compare with its competition in this area?

X. IMPROVEMENT

A. What is your overall opinion of the company's customer service?
B. Please rank your suppliers according to the quality of customer service they provide.
C. Do you have any suggestions as to how the company can improve its customer service?
D. How will these improvements help your operation?
E. Will they have any meaningful effect on the volume of business you do with the company?

B. POLICY CONSIDERATIONS RELATED TO CUSTOMER SERVICE AND INVENTORY MANAGEMENT

Before proceeding to develop methods of measuring customer service, consideration should be given to establishing corporate policies.

CASE IN POINT: A cutlery manufacturer found, for example, that its customer service level (defined as a percent of line items filled complete within 24 hours from receipt of order) improved 12% to 96.7% when it redefined back-orders to exclude special order and other items. This apparent "improvement" in customer service so motivated the sales force, that sales on regular items increased 3%.

The company's analysis during one month showed, for example, that 63% of the line items back-ordered were due to the Production department's not meeting its schedules. This was due to a variety of factors; i.e., unavailability of packaging materials, poor expediting, unauthorized labor strike in one of the manufacturing plants, etc. Another 2% of the back-orders were due to an imbalance of inventories between shipping points; i.e., some warehouses were overstocked while others ran out of goods.

Out-of-stock conditions due to factors such as those described above should be reflected as back-orders in a customer service analysis. But other back-orders were being reported which were beyond the control of manufacturing and distribution management. The firm's top management decided that these occurrences should not be included in the measure of customer service: viz,

- *Special Order Items:* These were items which were not kept in inventory. They were made to customer specification. But the "Customer Service Report" included as a back-order, any order which could not be shipped within 24 hours of its receipt in the Order department. It can be seen from this example that it may be desirable to have more than one definition or policy as to what constitutes a back-order condition.

- *Discontinued Items:* Because of lack of proper communications with the sales force, orders were frequently received for items no longer carried in stock. Until such time as the respective product code numbers were deleted from the computer file, such orders were counted as back-orders.
- *Seconds:* Several customers purchased defective merchandise on a regular, "as available" basis. When "seconds" were generated at the plant, they were placed in inventory and orders could be shipped. At other times, however, these orders had to be held for later shipments.

Shown below are a number of policy considerations related to both customer service and inventory management. Both of the latter items must be considered simultaneously if customer service is to be maximized.

1. Establish when air shipments and other expediting delivery methods are to be used. Expediting may avoid stockouts, but extra costs are usually incurred.
2. Determine which items should be carried in inventory and which should be made to order. Too often, more time, effort and inventory dollars are devoted to slow movers which account for a small percentage of dollar sales, than to the fast moving inventory items which account for the major portion of dollar sales. As a general rule, the firm should apply customer service measurements only to those items which are considered important to the overall sales effort. Many products are carried in inventory as convenience items. Giving the latter items as much emphasis as the former items, can result in dissipation of production and inventory control effort. Then all customer service usually suffers.
3. Clarify all departmental responsibilities related to inventory management and control.

CASE IN POINT: A consumer products manufacturer having several plants in the U.S. and abroad, found that on the average, 55% of its finished products were en route to its distribution centers and not, therefore, available for shipment to customers. After pinpointing expediting responsibility, in-transit inventories were maintained at 25% of the total. The inventory reduction achieved through this acceleration of goods through the pipeline amounted to $5 million.

4. Establish procedures, lead times and lines of communications in handling sales/marketing promotions.
5. Specify under what circumstances requests for rush orders will be processed. Rush orders can create added manufacturing, purchasing, materials handling and transportation costs. Processing rush orders can also upset schedules on regular items, thus inadvertently causing back-orders.
6. Determine desirability of establishing a system which will continuously measure order cycle time (discussed in detail in Section C, below).
7. During periods of short supply, consider making split shipments and partial shipments rather than back-orders, until complete shipments can be made.

8. Consider utilizing a separate post office box number and providing customers with special delivery mailers.

A thorough review of policy considerations in your company will surely result in a number of areas which require more specific guidelines than now exist. The above list is not designed to be all inclusive or applicable to all situations.

C. SERVICE CYCLE DETERMINATION

One of the first steps in determining your service cycle (the time from placement of orders by the customer to the moment he receives the merchandise) is to list all of the activities or functions through which orders pass. Shown below is a typical list:

1. U. S. Mail
2. Company Mail Dept.
3. Order Entry Dept.
   a. Credit check
   b. Inventory availability check
   c. Routing to appropriate distribution center
4. Electronic Data Processing
   a. Prepares shipping documentation
   b. Prepares shipping labels
   c. Prepares invoices
   d. Transmits shipping data to distribution centers
5. Distribution Center Operations
   a. Prepares worksheets for order selection
   b. Pulls orders from warehouse and transports to truck loading dock
   c. Loads trucks
6. Transport Merchandise from Distribution Center to Customer

Each of the above items utilizes time, and this is what we must measure. Having a measure of where you are will then enable you to evaluate the effect of employing different systems. An automotive parts supplier was, for example, able to reduce his service cycle time by one day through utilizing his computer to perform all of the functions listed under item "3" above. The company also saved $42,000 per year in labor costs as the computer displaced several clerical personnel in the latter functional areas. A record distributor trimmed one day off its service cycle by establishing a separate post office box and bypassing the corporate mailroom.

Having listed the appropriate activities, proceed to measure the time in days required to process each order through this "system." Determine in-transit time to customers through use of mail-back cards. Most customers will gladly cooperate. Record all respective times on a detail sheet (see Exhibit 1-2, page 34) and then summarize results over some time period, say three months (see Exhibit 1-3, page 35).

D. SERVICE LEVEL

Service level, expressed as a percentage, is the figure which is most commonly referred to when "service levels" are discussed. It may be expressed in the following ways:

1. % of Line Items Satisfied on Time
   Calculated by:

   $$\frac{\text{Number of Line Items Filled During Time Period}}{\text{Total Number of Line Items Processed During Time Period}}$$

2. % of Units Satisfied on Time
   Calculated by:

   $$\frac{\text{Number of Units Filled During Time Period}}{\text{Total Number of Units Processed During Time Period}}$$

3. % of Dollar Demand Satisfied on Time
   Calculated by:

   $$\frac{\text{Dollar Demand Filled During Time Period}}{\text{Total Dollar Demand Processed During Time Period}}$$

Exhibit 1-1 shows a typical statistical display of the data used to develop the above customer service level percentages.

This entire analysis assumes that an item is given a back-order status if no merchandise is available as the respective order is received. A separate analysis may then be made to determine how long each item remains a B/O before merchandise becomes available, and/or how many orders are cancelled.

E. ORDER COMPLETENESS

Order completeness, defined as the percentage of orders shipped complete, is calculated in the following manner:

$$\frac{\text{Number of Orders Shipped Complete During Time Period}}{\text{Total Number of Orders Processed During Time Period}}$$

WEEK ENDING APRIL 7

TOTAL BACK-ORDER LISTING

INVOICE NO.	LINES ORD	LINES SHIP	LINES B/O	UNITS ORD	UNITS SHIP	UNITS B/O	VALUE ORD	VALUE SHIP	PRODUCT NUMBER	QUANTITY B/O	$ VALUE B/O
9793	5	4	1	9500	9250	250	461.	447.	2050	250	14.
11559	6	5	1	15500	13000	2500	1134.	901.	2210	2500	232.
11402	15	14	1	61000	56000	5000	3181.	2974.	2240	5000	207.
10983	12	9	3	16000	12500	3500	992.	709.	2230	500	25.
									7974	1500	116.
									3883	1500	140.
11464	11	10	1	5500	5000	500			2794	500	
14633	25	24	1	50000	49500	500	3239.	3213.	5870	500	25.
15766	16	15	1	29350	28100	1250	2065.	2000.	4440	1250	64.
13401	21	20	1	131300	123800	7500	12242.	11854.	3231	7500	388.
13163	9	8	1	8545	7545	1000	639.	550.	6575	1000	88.
12818	8	7	1	30000	27500	2500	2755.	2626.	1272	2500	129.
22999	18	12	6	22750	18750	4000	1770.	1443.	2781	500	38.
									6734	500	35.
										250	14.
										500	35.
										2000	186.
										250	17.
22809	7	4	3	13250	8750	4500	872.	556.	7474	2000	140.
										1250	87.
										1250	87.
32001	5	4	1	9560	7060	2500	4231.	3998.	2335	2500	232.
512	8	7	1	17000	16000	1000	724.	687.	9772	1000	37.
36586	37	34	3	43840	42340	1500	3491.	3477.	5544	500	13.
										500	13.
										500	13.
675	15	13	3	31900	30400	1500	1571.	1571.	6873	1000	120.
										1500	105.
										500	18.
2298	41	38	3	61150	58650	2500	4191.	3470.	4429	2500	232.
TOTALS -	260	228	32	556145	514145	42000	43558	40476	--	43500	2950

Total Line items processed	W/E 4/7 =	3,200
Total No. of units processed	W/E 4/7 =	4,200,000
Total $ Demand processed	W/E 4/7 =	285,000

% Line Items Satisfied on Time $= \frac{3,200 - 32}{3,200} = \frac{3,168}{3,200} = 99\%$

% Units Satisfied on Time $= \frac{4,200,000 - 42,000}{4,200,000} = \frac{4,158,000}{4,200,000} = 99\%$

% $ Demand Satisfied on Time $= \frac{285,000 - 2}{285,000} = \frac{284,998}{285,000} = 99\%$

Exhibit 1-1

DAILY ORDER CYCLE RECORD

ACCOUNT NO.	3-DIGIT ZIP CODE LOCATION	TIME IN U.S. MAIL	CO. MAIL ROOM	ORDER ENTRY DEPT.	E.D.P.	DISTRIBUTION CENTER OPERATIONS	MERCHANDISE IN-TRANSIT TO CUSTOMER	TOTAL DAYS
23527	109	3	2	2	1	2	6	16
10113	331	5	2	1	1	2	5	16
32231	072	6	3	2	1	2	7	21
55071	117	5	2	2	1	3	6	19
17341	072	5	3	1	1	2	5	18

Exhibit 1-2

ORDER CYCLE STUDY

3 MONTHS

MAY-JUNE-JULY

3-DIGIT ZIP CODE LOCATION	NO. OF ORDERS IN STUDY	RANGE OF ORDER CYCLE IN DAYS	AVERAGE TIME IN DAYS: U.S. MAIL	CO. MAIL ROOM	ORDER ENTRY DEPT.	E.D.P.	DISTRIBUTION CENTER	MERCHANDISE IN-TRANSIT TO CUSTOMER
109	70	15 - 26	6	3	1	1	2	5
331	27	13 - 58	7	2	2	1	1	6
072	65	13 - 46	6	2	2	1	3	4
117	44	13 - 51	4	2	2	1	2	9
072	9	15 - 19	5	2	2	1	1	6
186	27	10 - 18	4	3	2	1	3	3
006	74	14 - 22	4	3	2	1	1	7
232	35	10 - 13	2	2	2	1	3	2
060	26	14 - 16	3	3	2	1	2	3
941	51	12 - 17	4	2	2	1	3	5
600	21	14 - 19	5	2	2	1	1	6
071	72	12 - 18	4	3	1	1	1	7

Exhibit 1-3

2

Organizing the Physical Distribution Function for Top Profits

The most up-to-date physical distribution organization structures, along with detailed, functional responsibilities, are presented below. This data will assist you in developing an efficient, profitable, customer-service-oriented organization structure. Also shown are a number of ideas which will be helpful in improving the quality and productivity of the warehouse work force.

The last three decades have seen management scientists producing volumes of literature on the subjects of manufacturing, accounting, sales and marketing management organizational structures. The result of these efforts is a fairly uniform approach to positioning the latter functional areas within the organization. The predominance of one of these functions over others in the highest echelons of the organizational structure usually reflects the orientation of the business; i.e., capital equipment manufacturers emphasize manufacturing management, while cosmetic firms stress the sales and marketing functions.

Within the last decade there has been a recognition of the fact that a number of activities, traditionally dispersed within different departments, are, in fact, concerned with one central theme; e.g., physical distribution. More recently, advanced business managers have perceived that physical distribution is part of an even more universal activity within companies; that is, materials management.

While the concept of a proper organizational structure for physical distribution and materials management has been discussed conceptually and some empirical studies have been conducted on these subjects, only scant guidelines may be found to assist top management and the warehouse manager in their efforts to determine what physical distribution activities should be placed in the organization and where they should go, to maximize customer service and profitability.

A further limiting factor in effecting positive organizational change in this regard is the problem of internal politics. As noted earlier, the specific functions comprising physical distribution and materials management have been dispersed among the established departments within the organization. It has been observed that some managers are loath to relinquish control over these functions. While top management may intuitively feel that centralized control over physical distribution functions would be more desirable, they find it difficult to change the established order.

Finally, as will be apparent in the discussion below, the hard fact is, that it is difficult to find the calibre of person who is capable of managing those functions comprising the physical distribution and/or materials management concept. Where this is true, top management has no support to implement a change in organizational structure even if other problems could be overcome. Where necessary, warehouse managers should update their background and skills in order to be aware of the tools available and necessary for the successful management of the modern P.D. organization.

What does all this mean to the warehouse manager? Too often, poor organizational structure results in his having a lot of responsibility with too little authority. He is, in that case, constantly reacting to problem situations rather than controlling his operation—viz, responsible for inventory control and customer service with no authority over order entry or inventory *management* functions; responsible for controlling and reducing costs without having industrial engineering or a financial staff; responsible for documenting and maintaining accuracy of transactions without having a systems analysis and control capability.

Proper organizational structure is the foundation upon which a balance between operational responsibility and authority can be built. Our purpose in the following pages is to clarify the nature of modern physical distribution functions and their relationship to the overall organizational structure. Putting specific pieces together in an order appropriate to individual situations and effectively staffing the organization remains a management function.

Numerous surveys of distribution firms have shown that those having an organization structure and departmental responsibility breakdown approaching those recommended later in this chapter, have experienced profit and customer service level improvement. The following are typical examples:

- A coarse paper and sanitary products distributor reported a 50% improvement in order cycle time and 20% reduction in inventories.
- A wholesale meat distributor was able to reduce clerical costs by $144,000 per year through job consolidation, as Order Entry, Customer Service, Inventory Control and Purchasing departments were brought together under Distribution Management. Project 6% defrayed wage increases over 10 years onto the latter savings and the amount becomes even more significant!

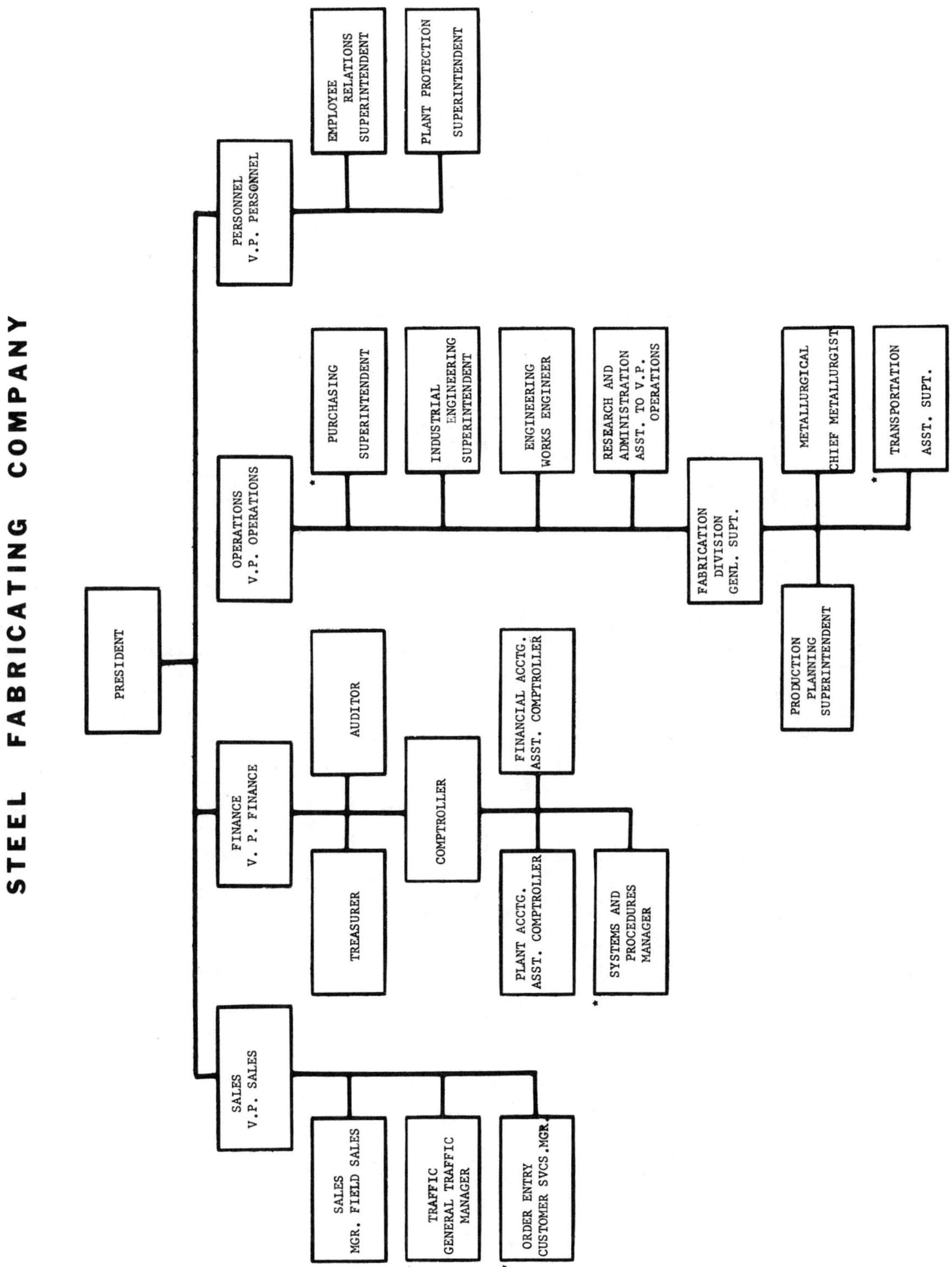
STEEL FABRICATING COMPANY
PRESIDENT
SALES V.P. SALES
SALES MGR. FIELD SALES
TRAFFIC GENERAL TRAFFIC MANAGER
ORDER ENTRY CUSTOMER SVCS.MGR.
FINANCE V. P. FINANCE
TREASURER
AUDITOR
COMPTROLLER
PLANT ACCTG. ASST. COMPTROLLER
FINANCIAL ACCTG. ASST. COMPTROLLER
SYSTEMS AND PROCEDURES MANAGER
OPERATIONS V.P. OPERATIONS
PURCHASING SUPERINTENDENT
INDUSTRIAL ENGINEERING SUPERINTENDENT
ENGINEERING WORKS ENGINEER
RESEARCH AND ADMINISTRATION ASST. TO V.P. OPERATIONS
FABRICATION DIVISION GENL. SUPT.
METALLURGICAL CHIEF METALLURGIST
TRANSPORTATION ASST. SUPT.
PRODUCTION PLANNING SUPERINTENDENT
PERSONNEL V.P. PERSONNEL
EMPLOYEE RELATIONS SUPERINTENDENT
PLANT PROTECTION SUPERINTENDENT

Exhibit 2-1

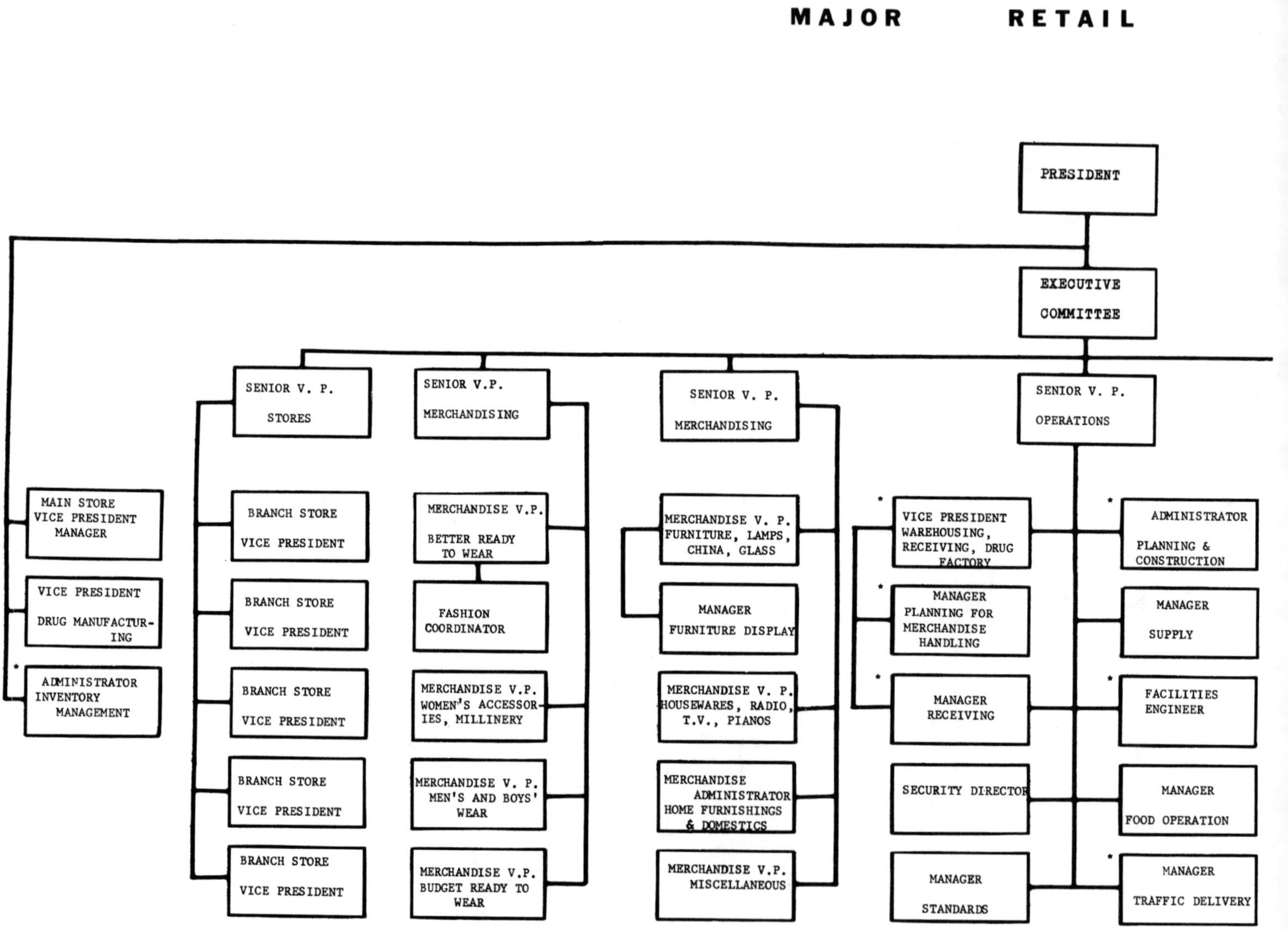
MAJOR RETAIL
PRESIDENT
EXECUTIVE COMMITTEE
SENIOR V. P. STORES
SENIOR V.P. MERCHANDISING
SENIOR V. P. MERCHANDISING
SENIOR V. P. OPERATIONS
MAIN STORE VICE PRESIDENT MANAGER
VICE PRESIDENT DRUG MANUFACTURING
* ADMINISTRATOR INVENTORY MANAGEMENT
BRANCH STORE VICE PRESIDENT
BRANCH STORE VICE PRESIDENT
BRANCH STORE VICE PRESIDENT
BRANCH STORE VICE PRESIDENT
BRANCH STORE VICE PRESIDENT
MERCHANDISE V.P. BETTER READY TO WEAR
FASHION COORDINATOR
MERCHANDISE V.P. WOMEN'S ACCESSORIES, MILLINERY
MERCHANDISE V. P. MEN'S AND BOYS' WEAR
MERCHANDISE V.P. BUDGET READY TO WEAR
MERCHANDISE V. P. FURNITURE, LAMPS, CHINA, GLASS
MANAGER FURNITURE DISPLAY
MERCHANDISE V. P. HOUSEWARES, RADIO, T.V., PIANOS
MERCHANDISE ADMINISTRATOR HOME FURNISHINGS & DOMESTICS
MERCHANDISE V.P. MISCELLANEOUS
* VICE PRESIDENT WAREHOUSING, RECEIVING, DRUG FACTORY
* MANAGER PLANNING FOR MERCHANDISE HANDLING
* MANAGER RECEIVING
SECURITY DIRECTOR
MANAGER STANDARDS
* ADMINISTRATOR PLANNING & CONSTRUCTION
MANAGER SUPPLY
* FACILITIES ENGINEER
MANAGER FOOD OPERATION
* MANAGER TRAFFIC DELIVERY

Exhibit 2-2

ORGANIZATION

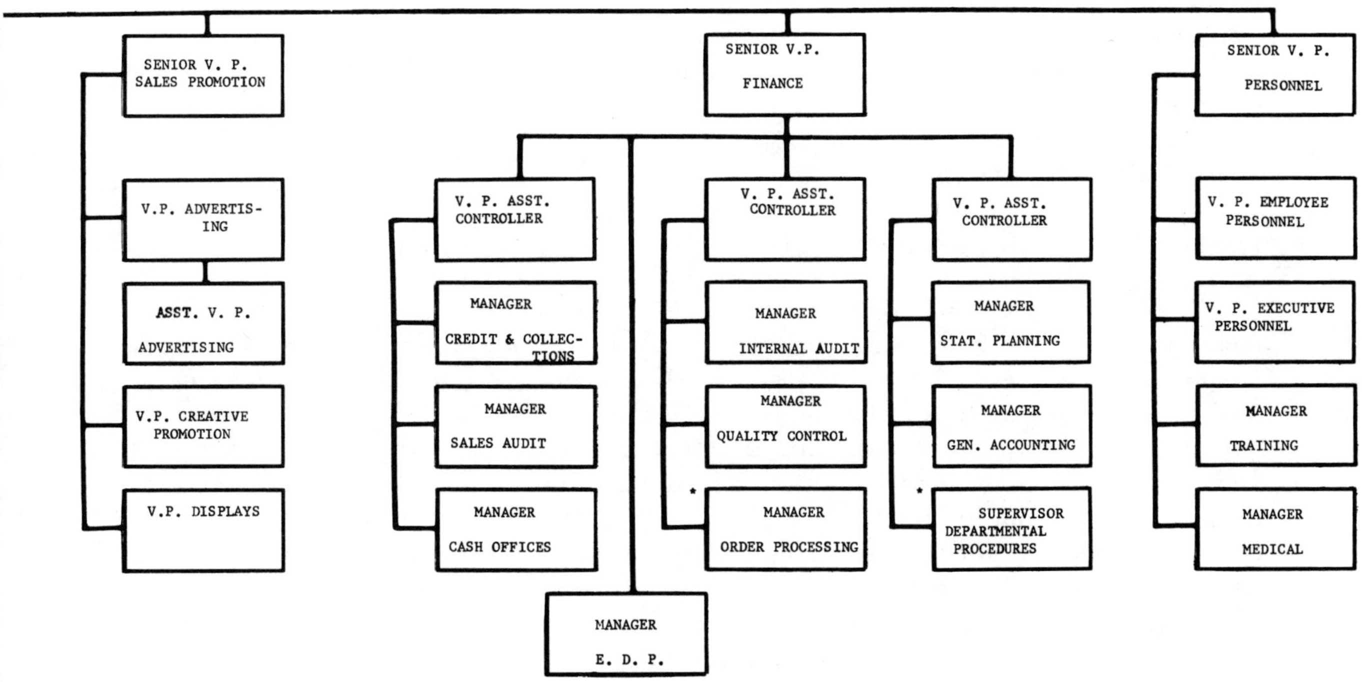

Exhibit 2-2 (Cont'd.)

- After reorganizing, a folding carton manufacturer saved $500,000 per year in avoidance of machine down-time. The company achieved better coordination between the inventory management, purchasing and production control functions. They saved 2,000 machine hours per year @ $250 per hour. Customer service improved also.

I. TRADITIONAL APPROACH TO ORGANIZATIONAL STRUCTURES

Shown in detail (Exhibit 2-1) is the organizational structure of a steel fabrication company. It represents what we shall call the traditional organizational structure insofar as physical distribution management is concerned.

Activities which should fall either entirely or in part under physical distribution are dispersed among the various functional departments. These are indicated with an asterisk. Other physical distribution functions exist within the company but are not felt by management to have sufficient importance to be included in their official organization chart.

Another "traditional" organizational structure is depicted in Exhibit 2-2. This organization chart represents the actual structure of a major retailer. Once again, an asterisk is shown next to those functions which fall, either in part or in whole, into the physical distribution category.

II. PHYSICAL DISTRIBUTION AND MATERIALS MANAGEMENT

The International Materials Management Society defines the Materials Management concept in the following terms: "Materials Management is a responsibility of executive management of any enterprise which has a tangible material and product flow. The MM objective is profit improvement and/or economic resource allocation. This is accomplished through the direction, coordination and evaluation of management skills and engineering techniques. It includes economic procurement and movement of material and product from supplier through processing and/or warehousing operations and on to the end user. Emphasis is placed on an efficient, highly competitive, cost-balanced service to the end user. The distinct skills directly involved in achieving this basic objective fall into two broad categories: Materials Management—Operations; Materials Management—Engineering. These functional categories must work together within a total systems concept, regardless of organization structure to accomplish the MM objective."

Physical Distribution usually suggests much narrower activity. Physical Distribution generally refers to those functions which are concerned with the movement of finished product from point of manufacture through all channels of distribution until it reaches the end user. It may also include the movement of raw materials to the beginning of the production process.

Physical Distribution is a more limited concept than Materials Management. The former has evolved within the framework of traditional organizational structures. Materials Management marshals all resources which are prerequisite to maximizing customer service and profitability. It does, however, require an organizational commitment on the part of top management.

First, let us review the diagrammatic representation of both the Physical Distribution and the Materials Management concepts (see Exhibit 2-3). Later in the chapter, specific line and staff functions are shown which may comprise Materials Management. Finally, suggested specific departmental responsibilities, both within and without the Materials Management area, are listed.

III. LINE AND STAFF ORGANIZATION FOR EFFECTIVE MATERIALS MANAGEMENT

A. Line Materials Management Structure

The line organizational structure shown in Exhibit 2-4 is most suitable where the company is divided into several autonomous divisions. Staff activities (see Exhibit 2-5) may then be most economically provided by a centralized, corporate group.

B. Line and Staff Activities Combined

The organizational structure shown in Exhibit 2-6 combines line and staff activities. It is the most recommended form where the firm's production and/or distribution operations are centrally controlled. This structure places the total distribution responsibility on one individual and gives him all the authority and skills required to meet this responsibility.

Staffing will depend on the size of the firm. In some instances, certain staff functions may be performed at significantly lower cost elsewhere within the organization. Also, as a practical matter, some managers may wear more than one hat.

Any practitioner in the distribution field knows that the above outlines of organizational structures represent only a part of the organizational question. There are other vital matters to consider; viz, the availability of capable people to staff the organization, the degree of cooperation and working relationships to be established with other groups in the company and the problem of how best to implement such a far-reaching organizational plan. All of these additional, practical considerations would be expected to influence the specific organizational form finally selected by individual firms. However, once agreement is reached on one of the foregoing organizational forms, modifications for practical reasons can be made most effectively.

C. Departmental Responsibilities Under the Materials Management Organizational Structure

The job of outlining a recommended organizational structure would be incomplete without delineating departmental responsibilities within the framework of the

SCOPE OF THE MATERIALS MANAGEMENT FUNCTION

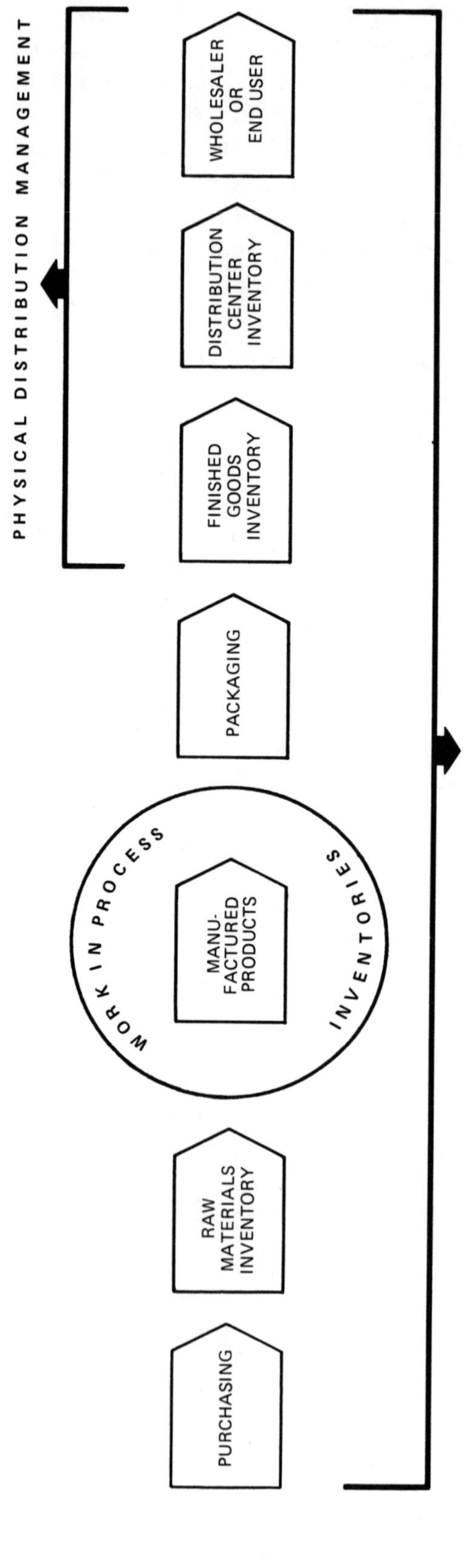

Exhibit 2-3

LINE MATERIALS MANAGEMENT ORGANIZATION STRUCTURE

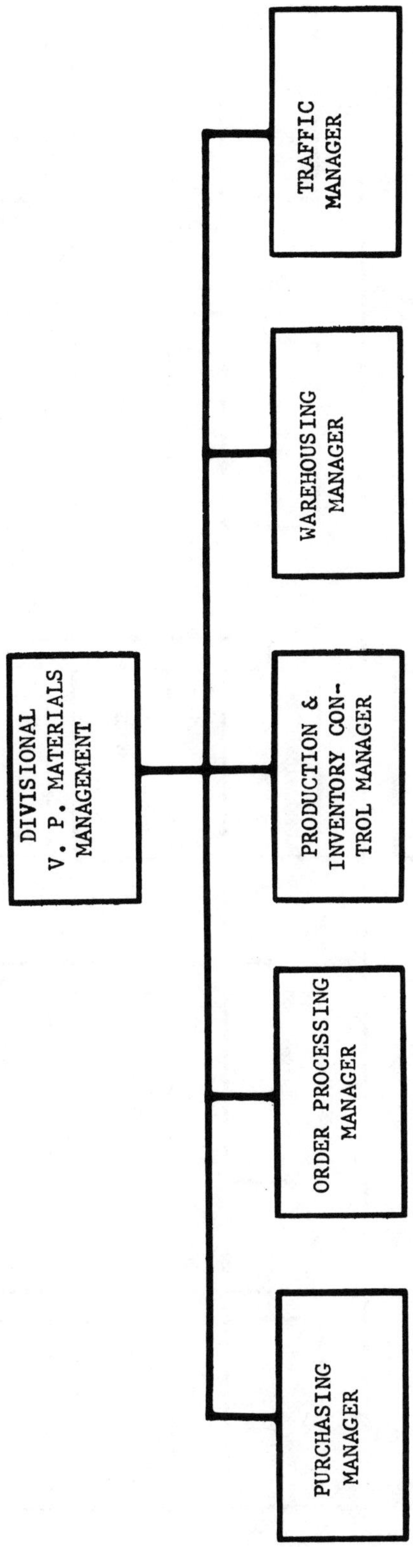

Exhibit 2-4

CENTRALIZED MATERIALS MANAGEMENT STAFF ORGANIZATIONAL STRUCTURE

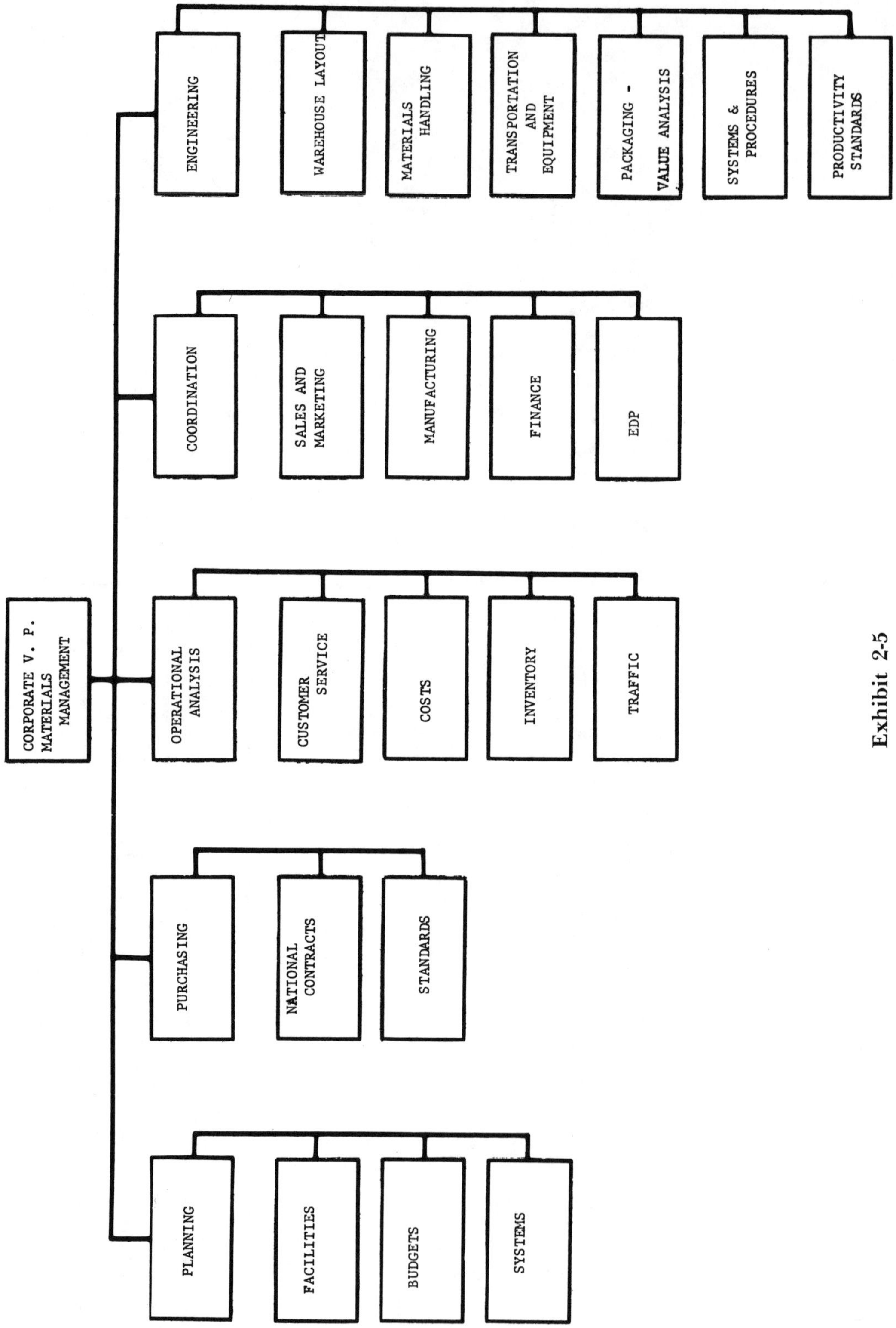

Exhibit 2-5

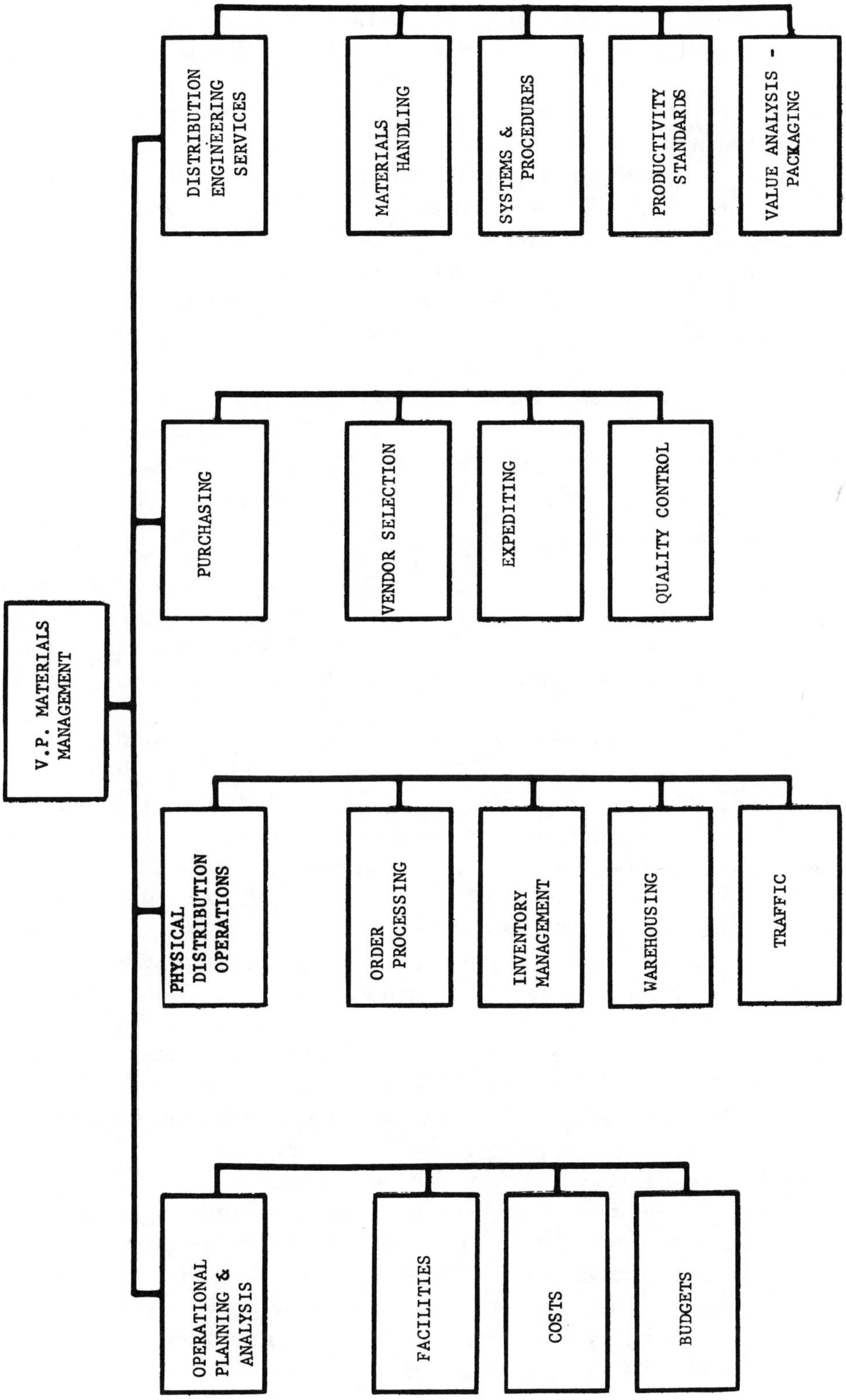
V.P. MATERIALS MANAGEMENT
OPERATIONAL PLANNING & ANALYSIS
FACILITIES
COSTS
BUDGETS
PHYSICAL DISTRIBUTION OPERATIONS
ORDER PROCESSING
INVENTORY MANAGEMENT
WAREHOUSING
TRAFFIC
PURCHASING
VENDOR SELECTION
EXPEDITING
QUALITY CONTROL
DISTRIBUTION ENGINEERING SERVICES
MATERIALS HANDLING
SYSTEMS & PROCEDURES
PRODUCTIVITY STANDARDS
VALUE ANALYSIS - PACKAGING

Exhibit 2-6

structure. We are concerned here with the practical, working relationships between materials management and the other, "traditional" major departments (see Exhibit 2-7).

Exhibit 2-8 shows the functional responsibilities defined within the Physical Distribution Department at J. C. Penney.

The Director of Distribution at J. C. Penney reports that his formation of an organizational structure (see Exhibit 2-9) along materials management lines has resulted in significant dollar savings to his company.

He also maintains that his reorganization from the former arrangement where P.D. functions were dispersed among several other departments, has enabled the company to expand rapidly and successfully.

Note the absence of an inventory management function at J. C. Penney. The company feels that in the retail business, inventory management should be placed at the store level. Local market conditions vary so greatly and change so rapidly, they say, that only the individual store buyers can effectively manage their inventories.

D. Personnel and Productivity

One of the most important single factors in the successful operation of any physical distribution function is the performance level of the work force. In spite of the fact that we have seen a rapid rise in the application of mechanization and automation in materials handling operations, productivity is still largely dependent on individual performance.

If a materials handling system is not functioning smoothly, the trouble can usually be traced to some part of the work force. For instance, if Receiving department personnel fail to count delivered goods, the effectiveness of the inventory control system is negated; moreover, a security problem is created since some truck drivers may be quick to take advantage of poor controls on the receiving dock by deliberately shorting orders. In addition, stockmen who misplace goods in storage cause unnecessary back-orders and additional purchases or manufacture of items; this ties up needed working capital as well as warehouse space. Order fillers who pull wrong items and shipping clerks who fail to catch their errors also contribute to a malfunctioning physical distribution "system."

It is also important to note that even the most sophisticated electronic data processing equipment in use today is not capable of detecting short counts and many other data entry errors. Data processing equipment and computers are magnificent tools for organizing information and expediting decisions. However, do not lose sight of the fact that the validity of figures is no better than their source.

Frequently, managers rationalize poor employee performance by saying, "You can't get good help in this tight labor market," or "What can you expect with the low wages that we pay our warehousemen?" While the above comments reflect real problems, it is important to ask if everything possible is being done to maximize the performance of our work force. That the main problem does not center around wage rates or tight labor markets may be seen in numerous studies which have been made of employee exit interviews, and in discussions with numerous materials handling managers representing a wide spectrum of industries.

For example, the bulk of employees who quit their jobs do so for a variety of reasons; wage rates rank low as a causative factor. More often employees complain about inability to work with their supervisors, or they are not given recognition for work well done. Further observation indicates that too often supervisors do not maximize their subordinates' performance. They are not, for instance, matching people and jobs, nor are they making a great enough effort to determine the actual performance level of personnel in their areas of responsibility. A distributor of sanitary products found, for example, that the introduction of a performance measurement system (actual against standard time) resulted in a 25% increase in materials handling productivity. The firm's direct labor force was reduced by 19 men. Average annual earnings of the latter group amounted to $6,500 including fringe benefits. Total annual savings —$123,500.

This section presents a series of building blocks which can assist the supervisor and manager in selecting and developing a better P-D work force. Also covered are several methods of improving and controlling employee performance. Implementing the suggested structure takes effort, continued attention and much individual initiative, but the results, in terms of lower employee turnover and a more efficient operation, are well worth it.

THE BUILDING BLOCKS OF INCREASED PRODUCTIVITY

1. JOB DESCRIPTIONS

The first foundation in our structure is a job description. The job description is simply a statement setting forth the characteristics, duties and responsibilities of a specific job. It covers such things as a general statement describing the job function; daily duties; periodic, monthly, weekly and occasional duties; superiors; subordinates; outside contacts; equipment used and working conditions.

The job description provides a blueprint of the job requirements against which an applicant can be measured. Too often, men are hired who soon quit or must be terminated because they are not suited to the job. A typical example of a poor man-job match is the receiving clerk who has no facility with basic arithmetic. The warehouse manager soon finds himself harassed by the Accounting department because of errors being created in the Receiving section. The controller of a textbook manufacturer found, for example, that several suppliers of raw material were shorting orders. It was discovered that the receiving clerk had great difficulty with arithmetic and therefore simply did not bother to make counts of incoming supplies. A reconstruction of receipts showed that approximate annual losses to the company amounted to $35,000. A simple test at the time of employment would have alerted warehouse management to this individual's deficiency. He could then have been placed in a position more in keeping with his abilities.

DEPARTMENTAL RESPONSIBILITIES UNDER PROPOSED MATERIALS MANAGEMENT ORGANIZATION STRUCTURE

SALES/MARKETING

1. Establish Desired Customer Service Level
2. Determine Disposition of Obsolete Material
3. Authorize Returns
4. Packaging Design
5. Promotion Schedules
6. Annual Sales Forecasts with Quarterly Update
7. Advisories of Anticipated Deviations from Historical Trend
8. Advisories of Special Sales Pressures

MATERIALS MANAGEMENT

1. Order Processing
2. Inventory Management at all levels; i.e., Raw Materials, Work in Process and Finished Goods
3. Warehouse Management
4. Traffic
5. Production Scheduling
6. Purchasing
7. Provide Desired Service Level

MANUFACTURING

1. Meet Production Quotas
2. Quality Control

FINANCE

1. Audit Distribution Operations
2. Establish Customer Credit Parameters

Exhibit 2-7

J.C. PENNEY

DISTRIBUTION DEPARTMENT FUNCTIONAL RESPONSIBILITIES

DISTRIBUTION/ MERCHANDISING COORDINATION	INDUSTRIAL ENGINEERING	PLANNING	RETAIL WAREHOUSING	TRAFFIC	WHOLESALE WAREHOUSING
Buyer and Field Guidance - Cost - Economic Order - Quantities - Landed Mark-up - Source Locations - Terms of Purchase - Pricing - Cost - Concessions	Facility Design	Research	Deliverable Merchandise Program	Rate and Route Analysis	Wholesale Warehouse Management
Volume Shipment Programs	Material Handling Systems and Equipment	Strategy	Plans and Coordinates all steps from location to opening of pool ware-houses	Consolidation	Order Processing
Alternate Distribution Methods	Equipment Specifications	Short and Long Range Plans	Customer Delivery	Supplier Shipping Instructions	Merchandise Storage
Supplier Services - Warehousing - Supplier Trucks - Pool cars	Stockroom Space Standards	Develop and test new Programs - Bolt - Catalog Replenishment for Small Stores - Freight Bill Payment	Outside Stockrooms	Carrier Service	Store Shipments
Furniture and Appliance Storage in Transit and Shipping Programs	Operating Methods and Procedures	Distribution Models - Sales - Suppliers - Transportation	Short term Space Requirements	Special Shipping Programs	Order Invoicing
	Economic Evaluation	Long Range Data Requirements	Cost Evaluation	Private Carriage	Liaison with Buyers
	Budgets	Long Range Systems	Training Programs	Shipping Violation Chargebacks	Warehousing Cost Standards and Control
	Cost Standard			New Store Merchandise Shipment Control	Warehouse Space Planning
	Control Systems			Claims	Merchandising Programs
	Operation Cost Improvement			Catalog and Media Mailing	
	Document Control			Postal Service	
				Government Legislation and Regulations	
				Agencies and Councils	

Exhibit 2-8

J. C. PENNEY DISTRIBUTION ORGANIZATION

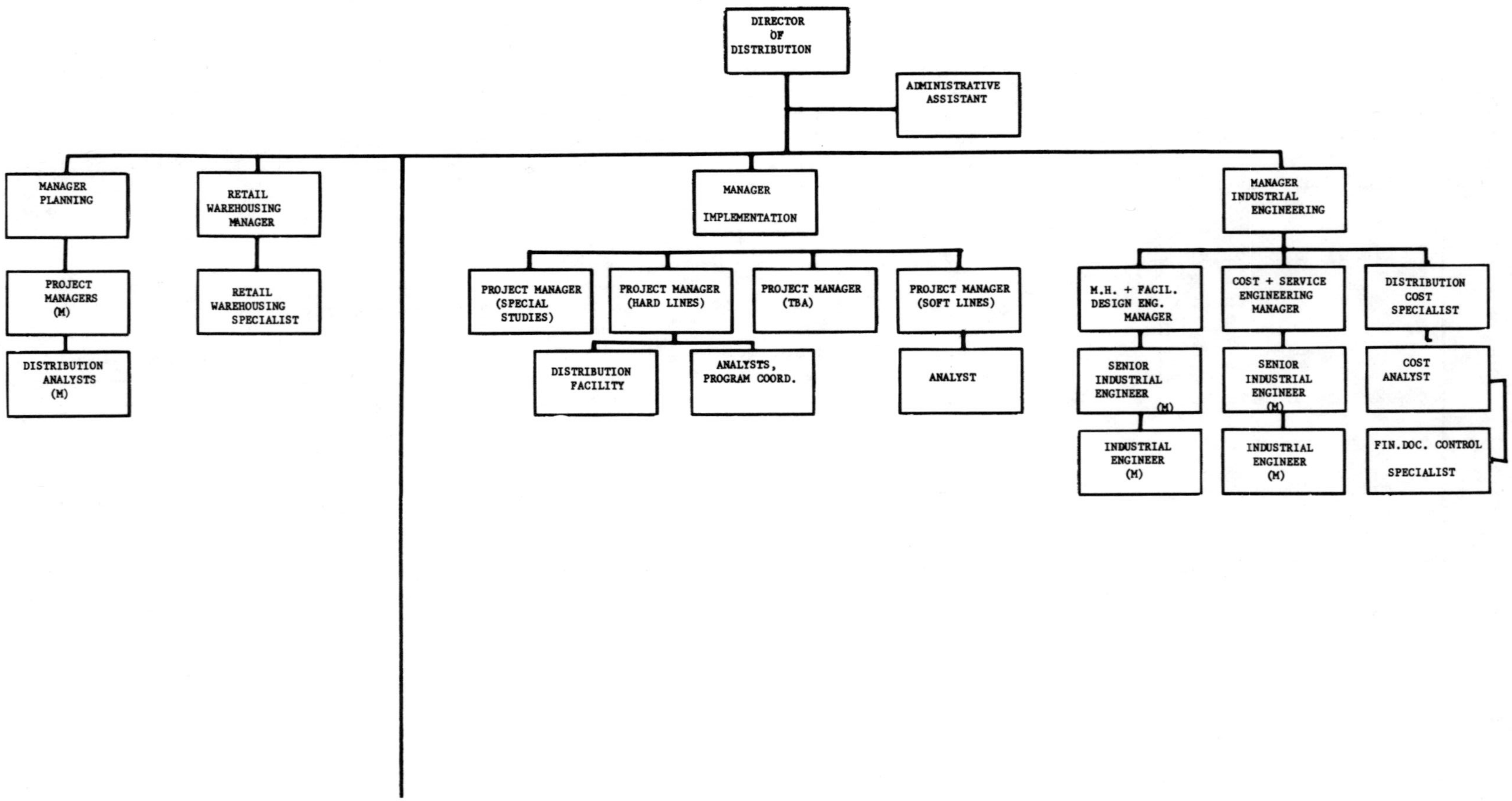

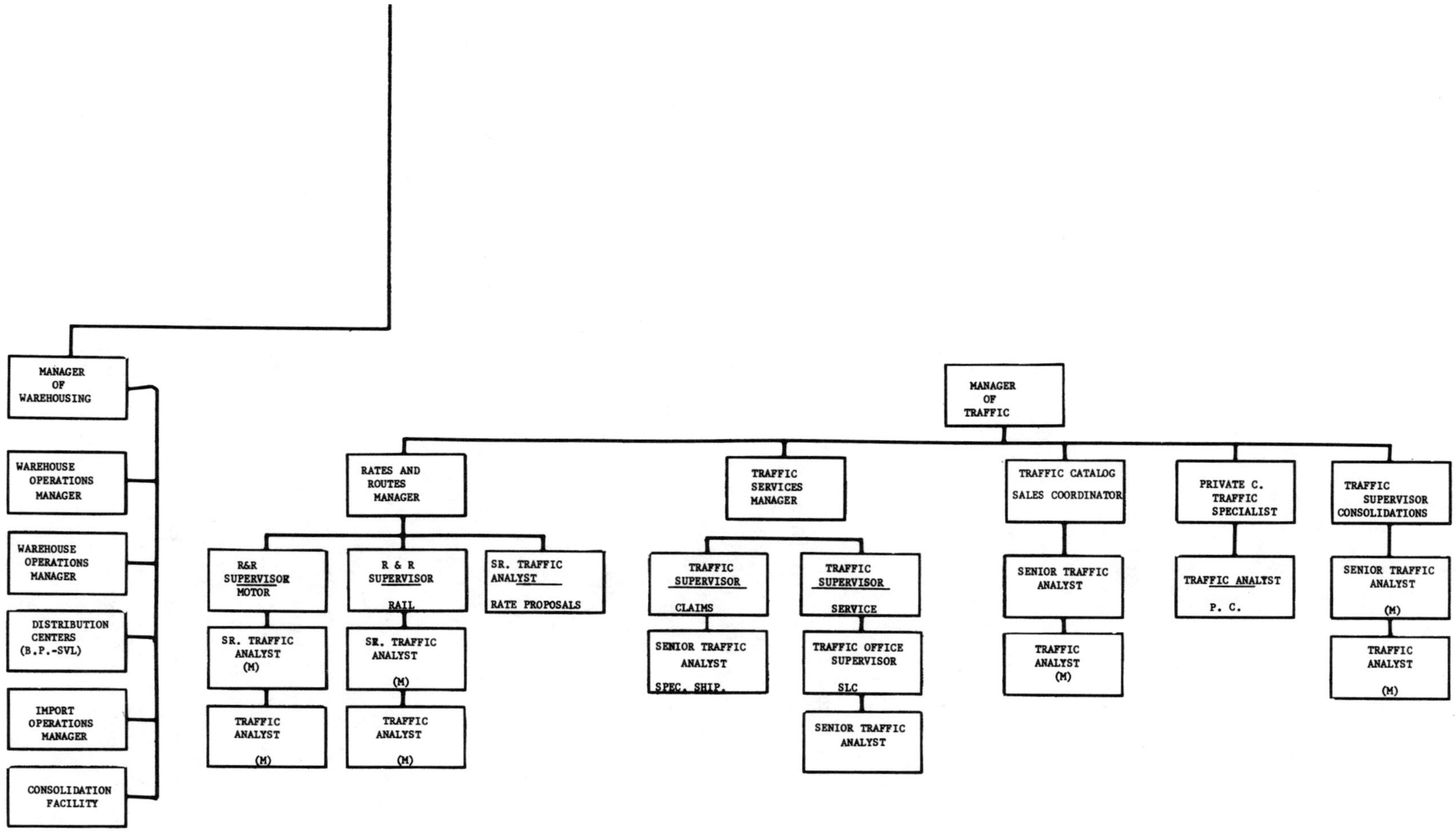
MANAGER OF WAREHOUSING
WAREHOUSE OPERATIONS MANAGER
WAREHOUSE OPERATIONS MANAGER
DISTRIBUTION CENTERS (B.P.-SVL)
IMPORT OPERATIONS MANAGER
CONSOLIDATION FACILITY
MANAGER OF TRAFFIC
RATES AND ROUTES MANAGER
R&R SUPERVISOR MOTOR
SR. TRAFFIC ANALYST (M)
TRAFFIC ANALYST (M)
R & R SUPERVISOR RAIL
SR. TRAFFIC ANALYST (M)
TRAFFIC ANALYST (M)
SR. TRAFFIC ANALYST RATE PROPOSALS
TRAFFIC SERVICES MANAGER
TRAFFIC SUPERVISOR CLAIMS
SENIOR TRAFFIC ANALYST SPEC. SHIP.
TRAFFIC SUPERVISOR SERVICE
TRAFFIC OFFICE SUPERVISOR SLC
SENIOR TRAFFIC ANALYST
TRAFFIC CATALOG SALES COORDINATOR
SENIOR TRAFFIC ANALYST
TRAFFIC ANALYST (M)
PRIVATE C. TRAFFIC SPECIALIST
TRAFFIC ANALYST P. C.
TRAFFIC SUPERVISOR CONSOLIDATIONS
SENIOR TRAFFIC ANALYST (M)
TRAFFIC ANALYST (M)

Exhibit 2-9

Obviously, the job description is merely the first step in preventing such situations from occurring. Not only must the job description be thoughtfully prepared but supervisors must also be trained to use them effectively; responsibility for follow-through must be pinpointed. Exhibits 2-10 through 2-13 show illustrations of job descriptions in the physical distribution area. (See pages 56-59.)

2. COMPREHENSIVE RECRUITING

This second foundation block is fundamental to the success of our entire system. The basic principle offered here is that you must have quantity in order to get quality. In a tight labor market such as we have today, this is not easy to achieve. In addition to the more usual recruiting sources, such as friends, associates, present employees, employment agencies, college and high school campuses, don't overlook trade associations, governmental agencies, chambers of commerce, social, fraternal and religious groups and selective advertising.

3. PRELIMINARY INTERVIEW

The preliminary interview is the first step in evaluating prospective employees, and it should be accomplished in the first 10 minutes or so of speaking with the candidate. This should be considered a screening process to eliminate those persons who do not warrant the further expenditure of time and money which the employment process requires. The preliminary interview is further recommended because it helps you to keep on a definitive path and to cover all the things that might be important, in accordance with the job description.

For instance, the job description might call for regular customer contact, as in the case of a delivery driver. In this situation, physical appearance and personality characteristics would be important considerations that can be quickly evaluated. Another characteristic that cannot be evaluated well from objective tests, and which requires a personal interview, is voice. If you were hiring a supervisor, it would certainly be necessary to make some kind of record of the impressions you gained from hearing him speak. A high-pitched, wavering voice that carried little sound of authority would weigh heavily against an individual applying for a supervisory position.

The ability to express oneself is frequently important and will be shown on good job descriptions for certain jobs. Does the individual get his ideas across fairly well, or is he somewhat scattered, confused and illogical? Or, on the other hand, is he logical, clear, convincing and superior in his ability to express himself? Does he seem timid or cocky, or is he reasonably self-assured?

In considering supervisory level personnel, it is most important that we select individuals who do not have the tendency to talk down to their subordinates. The most effective supervisors are those who can make their subordinates feel that they are part of the operation and those who are also receptive to suggestions.

If the applicant "passes" this preliminary interview, continue on to the next phase of the interviewing process.

4. PERSONAL HISTORY

In this part of the interview, a personal history inventory is developed. It should cover such things as general history, military service, experience, social history, etc. In gathering the above information, it is vital that we determine the following:

1. Does the applicant meet the physical and mental requirements indicated in the job description?
2. Does he show adequate social stability? A positive diagnosis would mean that he will get along well with his co-workers, not antagonize or alienate persons with whom he comes into contact.
3. Is he properly motivated for this particular job? If he is, he is likely to remain with the company as a long-term employee. The job should represent a challenge to him, but it should not be such that it is too far above his achievement level.

It is important to mention that the job description should be referred to constantly throughout this phase of the interview.

5. THE DIAGNOSTIC INTERVIEW

Now, having some facts to work with, and having judged that the applicant has some potential for the firm, it is time to conduct the diagnostic interview. This is an item that cannot be easily handled by a novice. If trained personnel are available within the organization, their talents should be utilized at this point; if not, serious consideration should be given to using a personnel consulting service.

The diagnostic interview covers many of the items provided in the personal history inventory, but an oral interview is now conducted where certain clarifications are made of the individual's background. For example, ask the individual why he liked or disliked a particular job; you may be able to judge from his response and reactions how well he would fit into the work environment that is being offered. You should know if the individual has experienced the feeling of having taken part and participated in group activities.

It is important to determine what his early ambitions were and how he has worked toward fulfilling those ambitions. Does his social history reveal that he has the capacity for learning new material that would prepare him for advancement? Also, evaluate the progress the individual has made toward maturing socially and emotionally. A check should also be made of the economic and financial history of the individual. It is important to learn whether his economic aims are in line with the opportunities that are offered, and whether the salary being proposed is in keeping with the applicant's requirements. It has been found that dishonesty is often caused by pressures resulting from employees living beyond their means. Add up a man's debts and expenses and see if these make him a risk. An air conditioner manufacturer found, for example, that a recently hired warehouse manager, working in collusion with truckers, misappropri-

G+W Consolidated Cigar Corporation

JOB DESCRIPTION

Job Title	Date:	Pay Grade:
SUPERVISOR - DISTRIBUTION CENTER	3/1/	III
	Location: Bristol	EDP Code: 104703
Approved by: G. Sutliff	Department: Allied Products	[X] Exempt [ ] Non-exempt

Job Summary

Under general direction, responsible for the efficient supervision of receiving, storage, packaging and shipping of imported and domestic allied products.

Job Duties

1. Supervises the taking of prescribed inventories and maintenance of inventory records.
2. Assures that incoming merchandise is properly stored in a predetermined area on a first in - first out basis.
3. Assigns employees to jobs according to their skill and classifications. Follows up work assignments to insure that prescribed instructions and procedures are properly adhered to by staff.
4. Reviews packaging material inventory and requisitions needed material as required to avoid delays in shipping to customers.
5. Coordinates scheduling of operations to maintain an adequate supply of various marketing counter display deals.
6. Supervises the repairing of damaged and/or defective lighters and pipes returned by customers. Maintains records of repairs and cash receipts.
7. Responsible for the picking, checking and packing of orders; routing and preparation of shipping documents; tracing strayed shipments. Prepares and maintains required records of shipment.

The above statements reflect the general details considered necessary to describe the principal functions of the job, and are not to be construed as a detailed description of all the work requirements that may be inherent in the position.

Exhibit 2-10

Consolidated Cigar Corporation

JOB DESCRIPTION

Job Title	Date:	Pay Grade:
Supervisor - Tobacco Warehouse	1/1/	III
Approved by:	Location	EDP Code:
S. Hernes	Bristol	100203
	Department:	[X] Exempt / [ ] Non-exempt

Job Summary

Supervises the unloading, warehousing and loading of all leaf and scrap tobacco. Responsible for the movement of all blended, threshed or leaf tobacco to and from each process. Supervises the maintenance of inventory records on all tobaccos stored in warehouse.

Job Duties

1. Receives all unblended leaf tobaccos. Supervises the checking of received load; fumigation and records and assigns storage space in warehouse. Stores imported tobacco as required by bonding regulations by lot numbers and U.S. customs entry numbers. Supervises the storage of bonded tobacco and insures that the bonded warehouse rules and regulations are maintained.

2. Supplies the required amounts and types of tobacco for the casing, curing and threshing.

3. Supplies Blending Department with the various types of tobacco required for each blend. Supervises the tobacco clerk in the determination of blends in short supply thereby setting a priority within the blending schedule. Stores all blended filler noting location stored for inventory records.

4. Receives orders from cigar plants for blended filler tobacco. Supervises the assembly of such orders and the preparation of the bill of lading and withdrawal notice. Supervises the loading of these shipments to cigar plants. Ships by-products as required.

5. Responsible for the fumigation of the entire plant.

6. Supervises the maintenance of all stored tobacco records.

The above statements reflect the general details considered necessary to describe the principal functions of the job, and are not to be construed as a detailed description of all the work requirements that may be inherent in the position.

Exhibit 2-11

G+W Consolidated Cigar Corporation

JOB DESCRIPTION

Job Title	Date:	Pay Grade:
Supervisor - Cigar Shipment	1/1/	V
	Location: Bristol	EDP Code: 101405
Approved by: S. Hernes	Department:	[X] Exempt [ ] Non-exempt

Job Summary

Assembles and ships cigars to customers as indicated by shipping orders.

Job Duties

1. Receives and stores in humidor finished cigars so as to facilitate quick and orderly assembly of customer shipments.

2. Receives shipping orders, consignee labels and humidor pick sheets by computer. Distributes pick sheets to humidor. (Pick sheets recap, by type of cigar, the finished products required for shipment for the day.) Supervises the loading and checking of stock withdrawn from humidor as indicated by the pick sheet insuring that stock is properly rotated by production dates.

3. Supervises the labeling of each carton of cigars and directs the storing of each carton by intended carrier. Assigns to each carton a consignee number for control purposes.

4. Supervises and maintains in process control records and checks labeled carton from shipping room to shipping platform.

5. Supervises the assembly of less than full carton customer orders.

6. Checks daily, the cigar inventory against orders shipped for that day. Reviews with supervisors inventories out of balance and attempts to correct such conditions.

7. Responsible for all finished product inventories. Spot checks inventories and completes physical inventory on a scheduled basis.

The above statements reflect the general details considered necessary to describe the principal functions of the job, and are not to be construed as a detailed description of all the work requirements that may be inherent in the position.

Exhibit 2-12

G+W Consolidated Cigar Corporation

JOB DESCRIPTION

Job Title	Date:	Pay Grade:
Production and Inventory Control Manager	June, 19	X
	Location: N.Y.O.	EDP Code: 1-125-10
Approved by: R. C. Phelan	Department: Operations	[X] Exempt [ ] Non-exempt

Job Summary

Under general direction, is responsible for developing production budgets, scheduling production and packing, and controlling production and inventory for all United States and Puerto Rico cigar manufacturing plants.

Job Duties

1. Plans and prepares production schedules for manufacture of products. Confers with appropriate department heads regarding availability of raw materials to coordinate periodic or seasonal sales requirements with production capacity and current inventory levels.

2. Prepares and issues operating instructions including basic cigar machine operating schedules and packing quotas to all cigar manufacturing plants.

3. Prepares and issues packing schedules for all special sales promotions and seasonal packing programs. Initiates changeovers of cigar machines from size to size, subject to requirements.

4. Maintains statistical records of operations and issues status reports to management regarding production, sales and inventory.

5. Prepares studies of operations including estimated output and periodic inventory levels.

6. Acts as liaison and coordinating agent between manufacturing plants and all other departments in matters pertaining to production.

7. Expedites operations that delay schedules and alters schedules to meet unforeseen conditions. Issues status and progress reports to other departments as required.

The above statements reflect the general details considered necessary to describe the principal functions of the job, and are not to be construed as a detailed description of all the work requirements that may be inherent in the position.

Exhibit 2-13

FIDELITY APPLICATION

AGENT'S NAME	COMPANY ☐ THE HOME INSURANCE COMPANY ☐ THE HOME INDEMNITY COMPANY

APPLICANT'S NAME IN FULL	AGE

ADDRESS (STREET, NO., CITY, STATE)

DATE OF BIRTH	PLACE OF BIRTH

NAME OF EMPLOYER REQUIRING BOND

POSITION OCCUPIED WITH EMPLOYER

NATURE OF EMPLOYER'S BUSINESS

GIVE EMPLOYER'S FULL ADDRESS

YOUR BUSINESS ADDRESS WILL BE:

AMOUNT OF BOND	EFFECTIVE DATE	PREMIUM TO BE PAID BY	LENGTH OF TIME IN SERVICE OF EMPLOYER REQUIRING THIS BOND.

SALARY RECEIVED	IF ON COMMISSION BASIS, STATE AVERAGE EARNINGS
PER	PER

ARE YOU	NUMBER OF PEOPLE DEPENDING ON YOU FOR SUPPORT
☐ MARRIED ☐ SINGLE ☐ DIVORCED	

YOUR REAL ESTATE VALUED AT	ENCUMBERED FOR	PERSONALTY VALUED AT	ENCUMBERED FOR
$	$	$	$

YOUR PERSONALTY CONSISTS OF:

OTHER DEBTS, (GIVE, WITH PARTICULARS, LIST OF DEBTS AS PRINCIPAL OR ENDORSER)

AMOUNT AND PARTICULARS OF ANY PRIVATE INCOME

IF EVER IN BUSINESS FOR YOURSELF, GIVE PERIODS WITH PARTICULARS AND WHY DISCONTINUED.

WERE YOU EVER BANKRUPT OR INSOLVENT?	GIVE PARTICULARS
☐ YES ☐ NO	

HAVE YOU EVER BEEN IN ARREARS OR DEFAULT IN A PREVIOUS EMPLOYMENT? (GIVE PARTICULARS ON SEPARATE SHEET)

☐ YES ☐ NO

IF EVER DISCHARGED FROM ANY POSITION, STATE PARTICULARS WITH DATES

IF YOU ARE NOW OR EVER HAVE BEEN BONDED, GIVE NAME OF EMPLOYER AND NAME OF SURETY COMPANY

IF EVER DECLINED A BOND, GIVE PARTICULARS

GIVE NAMES OF FOLLOWING: (IF PARENTS NOT LIVING, GIVE NAMES OF NEAREST LIVING RELATIVES)	FINANCIAL WORTH	RELATIONSHIP	P.O. ADDRESS
FATHER			
MOTHER			
WIFE OR HUSBAND			

Exhibit 2-14

If responsible for bad credit losses in above employment, or share in profits or losses of employers' business, give particulars

PREVIOUS EMPLOYMENT, ETC.

Give names and addresses of employers for past ten years, with position occupied, time engaged and reason for leaving each.

This data is important and replies must be full and complete. Be sure to give number of street in city addresses. Account for all your time during the past ten years whether employed or not. Attach separate sheet if space provided is insufficient for complete record.

Start with most recent employer

FROM WHAT DATE?		TO WHAT DATE?		NAME AND ADDRESS OF EMPLOYER	Position You Occupied and Where Located	Name and Present Address of Party Under Whom You Worked	Reason for Leaving
Month	Year	Month	Year				

NAMES OF OTHER REFERENCES	OCCUPATION	P. O. ADDRESS

I hereby agree for myself, my heirs, executors and administrators, in consideration of ☐ THE INSURANCE COMPANY ☐ THE INDEMNITY COMPANY, hereinafter called Company, becoming surety for me, and issuing the Bond of Security hereby applied for, or any renewal thereof, or any further or other Bond or Security on my behalf, to protect and indemnify the said Company against any loss, damage costs, counsel fees and expense of whatsoever kind that it may sustain or become liable for in consequence of such suretyship. I also hereby agree that the vouchers, or other evidence of payment of such loss by said Company, together with vouchers or other evidence of payment of costs and expenses incurred in adjusting said loss, shall be taken as conclusive evidence against me and my estate of the extent of my liability to the said Company. I hereby further agree that the Company shall have the right to decline to grant the bond applied for; that, in case the bond is granted, the Company shall have the right to withdraw or cancel the bond at any time; and that the Company shall not be responsible for any loss or damage that I may suffer by reason of any such action.

I hereby authorize and request any and all of my former employers and others to furnish the Company a complete history of my services with them, together with any information they may have concerning my personal character, habits, ability, disposition, etc., and particularly a statement of the cause of the termination of my employment, hereby releasing them from any and all liability for damages of whatsoever nature on account of furnishing the requested information.

Dated and signed at ____________________

this ____________ day of ____________ 19____

____________________ Witness.

____________________ Signature of Applicant.

Exhibit 2-14 (Cont'd.)

ated finished goods valued at $65,000. Following this disclosure, a review of his personal assets and financial situation revealed that the warehouse manager owned a heavily mortgaged $70,000 home; was supporting three children who were going to private out-of-town colleges; owned three expensive cars and was otherwise in debt.

Serious consideration should be given to utilizing a fidelity bonding company for coverage of all warehousing personnel. They make thorough background checks for you. If you do not use a bonding company, it is strongly recommended that a typical fidelity bond application (see Exhibit 2-14) be used as part of the interviewing procedure. Experience has shown that many individuals walk away from the Personnel department reception area when faced with having to fill out such a document.

6. WORK REFERENCE INVESTIGATION

One of the best sources of information about an applicant is his former employer; however, the usual means of securing such information—that is, by letter—is usually very unsatisfactory. Letters of recommendation and similar written material rarely contain negative statements. They do not tell the whole story. The cost of a long-distance telephone call may be repaid many times over by the strategic information obtained about an employee, information which might not be so readily obtained from a written letter.

The most preferred method would be a personal meeting with the individual's former employer. At such meetings, visual expressions are often more revealing than the answers to the questions themselves. However, the job level being considered may not justify the time expenditure required by a personal meeting. Whichever method is followed, it is vital that your queries include one important question: "Would you rehire him?" A former employer might answer many questions in a positive manner, but experience has shown that when you pin him down and ask, "Would you rehire this man back again if you had a similar position open?", he is likely to give you a frank reply.

7. TESTING

Two main purposes are served by introducing aptitude tests to the interviewing process. First, such tests provide an effective, objective check on the interviewer's judgment, and second, they almost invariably provide information relative to the inner man. This information may not be clearly revealed in the earlier phases of the interviewing process but is vital in ferreting out potential failures and individuals having personal problems which might affect their performance.

Testing programs should, of course, be designed to suit the job level as well as the job description. That is to say, we would expect different tests for clerical positions as compared with electricians, and different tests for a warehouse manager as compared with a filing clerk. Naturally, there is some overlap in tests; we can generalize by saying that three basic areas should be covered by testing:

1. *Motivation*

   This section includes such factors as an individual's interests, financial goals and requirements. If the job's rate of pay falls below the applicant's needs, you can be certain that if he accepts employment, it will only last until he finds a position having a higher rate of pay.

2. *Control*

   The second area in which considerable time should be spent is control or temperament. Here, tests measure how the man will react to problems arising in his daily contacts, and to the duties the job description shows will be required of him.

3. *Abilities and Aptitude*

   For most jobs, it is important to know how well a man thinks, how quickly and thoroughly he can use what he has learned in the past when confronted with new problem situations. What is necessary to determine is the individual's capacities. It is important to know how well a man will organize the duties of his job, his ability to handle details and his ability to delegate work to others. A measure of a man's social intelligence is also of interest; i.e., his tact and diplomacy, his ability to say the right thing and influence other persons toward a desired direction. In other words, this is a measure of the person's insight into human nature.

8. POST EMPLOYMENT PROCEDURES

Industrial psychologists have pointed out the fact that very few employees think of themselves as working for a company. Rather, they work for individual supervisors. If supervisors are permitted, by their attitudes and actions, to corrupt younger members of the organization, productivity will be minimized. A breakdown of morale and discipline will occur. Lax management results in employee indifference. To counter this, an atmosphere of alertness must be created which demonstrates that management cares.

Specifically, what can the man in charge do? He can:

- Set realistic standards of performance. To be responsible for a quota, goal or budget while being denied the means of achieving these ends, leaves a man with two alternatives—to fail or to cheat. Standards of performance are not merely a rigid guide established from accounting or engineering data, but are an estimate of human capabilities.

CASE IN POINT: A food products manufacturer recently instituted a "Management by Objectives" system. Each manager was required to develop a set of goals for the following year which both he and his superior agreed to. Progress toward goal was evaluated each six months.

In the first year of the above program, $2.5 million in cost reduction projects were identified and the savings achieved.

- Maintain effective communications. Efficiency depends to a great extent on how employees react to the rules, procedures and general conditions under which they work. Good communications are essential if management is to discover the cause of low standards of performance, poor morale and inefficiency.

CASE IN POINT: A distributor of chemical supplies implemented a series of standard labor cost control reports which compared standard hours against actual time spent in all direct and indirect labor classifications. The reports quickly showed that several departments were operating inefficiently. But supervisors did not know how to improve matters.

The director of distribution held a series of weekly meetings with all supervisors in charge of cost centers. They discussed the reasons for negative variances and made suggestions for improvement. One immediate benefit of these meetings was the recognition that each cost center had excess personnel to cover for absenteeism. The group agreed to create a single "labor pool" from which they could all draw on as needed. This resulted in $75,000 savings per year.

- Follow through on performance. Periodic unannounced checks on employee activity should be made at all levels. And most important, employees should know that these checks are a normal part of management control. Knowing that their work is being monitored will deter them from yielding to temptations to take shortcuts, and will provide an incentive for improving performance.

3

Coordinating Market and Financial Interests with Physical Distribution Concerns

This chapter presents several tested ways in which the physical distribution manager can take the initiative in improving interaction with marketing and show profit and customer service improvement in the process.

Also described are the basic reports and types of financial and operating analyses which can be used to improve P-D planning and reduce costs. The final section of the chapter shows specifically how several companies used these procedures to make significant reductions in P-D operating costs—valuable ideas which can be applied to your business.

I. MARKETING CONSIDERATIONS

In recent years, the managements of many firms have recognized the need to better dovetail marketing and distribution functions for more profitable results. Several factors have contributed to this realization.

A. Product Proliferation

One of the most significant factors has been what marketing executives have termed "product proliferation." Refrigerators, for example, used to be produced in a limited range of sizes, all in white. Today, refrigerators are produced in various freezer combinations; many longitudinal and horizontal sizes are available in decorator colors. Product proliferation has increased inventory carrying costs and increased the probability of stockouts. Inventory carrying costs place a heavy burden on the financial resources of the company. Stockouts can represent lost sales as customers select substitutes. To counteract the financial and customer service pressures brought about by product proliferation, distribution management has developed sophisticated management systems. These systems will be treated elsewhere in the book.

B. Price/Cost Squeeze

The last two decades have produced what many management executives and economists have termed the "cost/price squeeze." This refers to the fact that a tendency has developed whereby a firm's costs increase more rapidly than the firm's ability to increase prices. Increased competition, availability of substitute products, changes in consumer tastes resulting in horizontal or declining product demand curves and other factors, can make it difficult for a firm to increase prices without suffering a loss in sales.

As a consequence of the cost/price squeeze, many firms increased advertising budgets with the frequent result that profit margins deteriorated further. A reduction in manufacturing costs will ease this problem and should be attempted. It should be recognized, however, that manufacturing cost reduction efforts have been underway for many decades. Savings opportunities here can, therefore, be relatively small, particularly in traditionally well-managed companies. Cost reduction and control in distribution operations, on the other hand, represents a fertile area. Exhibit 3-1 illustrates the profit leverage resulting from a reduction in physical distribution costs. If, for example, the firm's after-tax profit is 2%, a savings of $20,000 in distribution costs yields as much profit to the company as $1 million in sales.

C. Greater Customer Demands on Suppliers

As a result of increasing costs of their own, retailers and jobbers are demanding an ever increasingly high level of service from their suppliers. Competitive pressures require that the distribution firm ship small orders more frequently with concomitant higher packaging, warehousing, inventory and transportation costs. The distribution firm is compelled to either improve operating methods or suffer the erosion of its profit position. A physical distribution operation which is unresponsive to increased customer demands compromises the efforts of the Sales/Marketing departments. The astute marketing manager realizes the value of a physical distribution organization that can provide a higher-than-industry-average level of customer service. In this case, physical distribution becomes a sales tool.

One firm, a hardware manufacturer, found that a 50% reduction in order processing time from five days to two and one-half days, precluded the necessity to build certain regional warehouses, and thereby saved a $3 million capital investment. Automated credit checking, updating accounts receivables, preparation of all shipping

PROFIT LEVERAGE PROVIDED BY PHYSICAL DISTRIBUTION COST REDUCTION.........

If net profit on the sales dollar is 2.0% then....

A SAVING OF	IS EQUIVALENT TO A SALES INCREASE OF
.02	$ 1.00
$ 2.00	$ 100.00
$ 200.00	$ 10,000.00
$ 2,000.00	$ 100,000.00
$ 20,000.00	$ 1,000,000.00

Exhibit 3-1

documentation and transmission of orders to shipping locations resulted in the shortened order processing time span.

D. Supplier-Customer Relations

Retailers and jobbers are grappling with physical distribution costs of their own. Some manufacturers and distributors provide innovative distribution services to their customers in order to gain a competitive advantage.

CASE IN POINT: A major paper company has, for example, led the industry in its relations with paper merchants by conducting a comprehensive physical distribution survey in their interests. The survey covered a detailed financial analysis of each participating paper merchant along those outlines presented later in this chapter. The survey also included an analysis of space utilization and materials handling applications. The results were made available to all participants, whose identities were coded. All firms were thus able to compare their operations against others in their specific industry without compromising confidential company data and positions.

CASE IN POINT: One of the largest manufacturers of sealing tape had its packaging engineers visit its distributors to determine if these "users" had recommendations to offer. The company was startled to find the number of significant recommendations made to improve product identification and usability; viz, larger labels and numerals, label color coding, standardization of pallet sizes, increased caliper of corrugated cartons to prevent "leaning" in warehouses, shrink wrapping of pallet loads to reduce pilferage and much more. The company's marketing management utilized physical distribution improvements to improve its market position. Other firms have retained consultants to present physical distribution seminars for their jobbers' personnel.

E. Summary

As greater financial and competitive pressures combine to reduce profit margins, marketing and distribution management must combine their resources and talents to the company's advantage.

Physical distribution represents a real and significant cost of doing business. An inefficient physical distribution operation will be reflected in higher prices of the company's product. Cost reductions in the P-D area can result in a more competitive price, or the savings can be channeled into marketing, sales or other areas of the business.

The items listed below are viewed as being the major areas of marketing and physical distribution interaction:

1. *Customer Service Policies.* Analysis of the physical distribution process makes it possible to class customers by cost of service, or profitability, and, in turn, set realistic policies of customer service. The latter policy area is probably the single biggest problem existing today between marketing and physical distribution.
2. *Product Design and Nomenclature.* A product design which sells well is not necessarily a product design which moves well through the physical distribution system. Odd shapes which require oversize packages increase cube requirements and push up costs for handling, storage and transportation, as well as for the package itself.
3. *Market Forecasting.* Demands on the distribution system must be known in advance if the system is to function economically and well. Accurate forecasting of requirements becomes particularly critical to such activities as planning warehouse facilities and negotiation volume transportation rates.
4. *Number and Location of Warehouses.* Another aspect of customer service is involved here. It is an area where physical distribution's analytical skills can improve both marketing efficiency and profitability.
5. *Inventory Policies.* Against the cost and obsolescence risks of large inventories, Marketing must weigh the stockout and customer dissatisfaction risks of small ones. Physical distribution techniques aid in this decision by pinpointing costs associated with physical location of inventories, reorder points and order quantities.

6. *Pricing.* The Physical Distribution department finds itself in an extremely difficult position when pricing strategy does not take into account the real physical distribution costs that will be incurred, or when customer service levels will have to be lowered in order to show a profit.
7. *Order Processing.* Marketing's responsibility for sales review and credit check can be handled by appropriate decision and exceptions rules determined by Marketing, but there is compelling logic for making order processing part of the physical distribution responsibility.
8. *Channels of Distribution.* Physical distribution practices frequently parallel the changes in title or responsibility that take place through the sometimes intricate channels of marketing, sometimes involving as many as 40-plus handlings of product between first seller and ultimate buyer. Joint examination of these practices by Marketing and Physical Distribution will uncover many opportunities for reducing cost without impairing marketing efficiency.
9. *Trade Practices.* Many trade practices relating to units of sale—i.e., by weight, volume or number of units—impose costly requirements on the physical distribution system and in many cases no longer serve a useful purpose. One such example is the barriers to effective cost sharing by buyer and seller which take the form of trade practices. Joint study by Marketing and Physical Distribution is already overdue.

II. NEED FOR FINANCIAL ANALYSIS

Successful management planning and control of operations in both manufacturing and distribution companies requires a comprehensive knowledge of the firm's position. Such knowledge can be obtained as a by-product of the daily input of information extracted from sales orders, purchase orders, customer payments and various internal calculations. Physical Distribution, in its strategically important position, can provide data necessary for meaningful management reports and analysis in addition to the general accounting output package (see Exhibit 3-2).

Outlined below is a basic set of reports and types of analyses which the distribution firm should employ to indicate the strengths and weaknesses of the company, as a basis for corrective action and future planning.

III. A SUGGESTED LIST OF COMPUTER-GENERATED BASIC MANAGEMENT REPORTS FOR DISTRIBUTION FIRMS

A. Accounts Receivable

1. Invoicing.
2. Monthly statements.
3. Aged accounts receivable analysis.
4. Cash flow projection.

THREE WAYS OF TREATING DISTRIBUTION COSTS

	Small Order (20 pcs.)	Large Order (200 pcs.)
Sales	$20	$200
Cost of Goods Sold	10	100
Contribution to {distribution costs, overhead, profit}	$10	$100

1. Lump distribution costs with overhead.

	Small Order (20 pcs.)	Large Order (200 pcs.)
Sales	$20	$200
Cost of Goods Sold	10	100
	$10	$100
Distribution costs	2	20
Contribution to {overhead, profit}	$ 8	$ 80

2. Treat distribution costs as indirect and average them.

	Small Order (20 pcs.)	Large Order (200 pcs.)
Sales	$ 20	$200
Cost of goods sold	10	100
Distribution costs		
Warehousing		
Order processing	1.50	1.50
Order filling	.60	1.60
Checking	.50	.80
Packing	1.30	2.20
Transportation (both are "minimum" shipments)	4.50	4.50
Total	$8.40	$10.60
Contribution to {Overhead, profit}	1.60	89.40
	8%	45%

3. Treat distribution costs as direct.

Note that when marketing and physical distribution costs are properly treated as "direct" there is a significant difference in the rate of profitability on small orders vs. large. In the illustration, the contribution to overhead and profit is 45 percent on the large order, and only 8 percent on the small.

Exhibit 3-2

B. Accounts Payable

1. Cash requirements forecasting by schedule date.
2. Vendor purchase order control and performance reports.

C. Sales Statistics

1. Monthly sales by product line and/or by item.
2. Monthly sales by product line and/or by customer.
3. Periodic shipping analysis.

D. Inventory Analysis

1. Monthly stock status.
2. Monthly joint replenishment orders by item by warehouse.
3. Monthly "no use" item reports.
4. Weekly expedite reports on low stock items.
5. Total average working inventory report.
6. Monthly turnover report.
7. Lost sales and back-order analysis.
8. Physical inventory report.

E. Other

1. Product line, item, customer profitability analysis.
2. Financial statements.
3. Labor productivity control reports.
4. Stratification of items in stock as a percentage contribution to total sales/ inventory value.

F. Exception Reports

1. Accounts 30-60-90 days in arrears.
2. Accounts exceeding credit limits.
3. Below quota labor productivity.
4. New items—inventory/sales performance.

IV. TYPES OF ANALYSES

A. Sales Analysis and Cost Analysis

1. Achieving a satisfactory level of profitability depends in large measure upon the proper direction of sales effort, which, in turn, requires factual information that can be obtained only through analyses of actual sales as reported by warehousing operations.

By "analysis" we mean a breakdown or distribution of some total into component parts. In scientifically managed distribution firms, this breakdown or analysis of sales is made in many ways including the following:

A. *Analysis by merchandise departments.*

1) To show which lines of merchandise produce the greatest volume and which the smallest.
2) Analysis may lead to greater concentration of sales effort on goods with high average line extension and reasonable volume.
3) Analysis may lead to more aggressive promotion of high margin items.

B. *Analysis by each major supplier's line.*

1) To consider possible dropping of low volume parallel lines.
2) To permit a reduction in inventory carried to support low volume sales of a particular line of goods.
3) Analysis may lead to more aggressive promotion of high margin items.

C. *Analysis by salesman.*

1) To evaluate by quotas.
2) As a basis for revising territories.

D. *Analysis by type of customer and size of order.*

1) To consider elimination of customers who contribute too small an amount of trade.
2) To regulate frequency of calls.
3) To determine those customers who offer possibilities for more intensive cultivation.
4) To adopt policies designed to discourage small orders (refused credit, forced to pick up goods at warehouse, required to pay additional service charges, charged higher prices through adoption of quantity discount schedule).

2. In addition to the several types of sales analyses described above, most distribution firms employ a number of *cost analyses.* We shall separate these into analysis by *kind of cost* and analysis by *manner of application.*

A. Analysis by kind of cost is a comparison of expense items in dollars expressed as a percentage of net sales in relation to:

1) Previous years.
2) Other distribution firms in the same industry.

B. Analysis by manner of application utilizes the same procedure as described in "A" above. It covers the following:

1) By order size.
2) By customer.
3) By product line.
4) By territory.

C. In making a cost and net profit analysis by territory:

Step 1: Determine sales for each territory.

Step 2: Determine cost of goods sold for each territory.

Step 3: Compute gross profit for each territory by subtracting cost of goods sold from net sales.

Step 4: Charge each direct cost to the territory by subtracting cost of goods sold from net sales.

Step 5: Determine a suitable basis for allocating semi-direct costs and distribute them to the appropriate territories. Semi-direct costs are those expense items which cannot be charged directly to a territory, but a reasonable basis for apportionment can be established. Included in this category would be such items as delivery expense and billing expense.

Step 6: Charge each territory its proportionate share of indirect costs. Indirect costs are those expense items which have no measurable relationship with the territories. Examples would be office rent, legal expenses and executive salaries.

Step 7: Subtract all costs from gross profit to determine net profit for each territory.

B. Ratio Analysis to Reveal Overall Position

Through their combined review of ratio analysis, physical distribution, marketing and financial executives can determine the strengths and weaknesses of their firm in relation to others in their industry. Coordinated plans may then be developed to correct weak areas and improve strengths. There are two nationally recognized, published sources of ratio analysis; i.e., "Annual Statement Studies," published by Robert Morris Associates, and "Key Business Ratios," published by Dun and Bradstreet, Inc. Both of these firms provide a broad range of comparative ratio analyses *by type of industry*. Meaningful comparisons can thereby be made. See Exhibit 3-3.

C. How to Analyze Your Distribution Costs to Achieve Greater Profits

Assume for the moment that a fictitious firm called Soaring Satellite, Inc. purchased $25,000 worth of products from your company. Your overall average profit on sales equals 10%. Does this mean that your company made money on Soaring Satellite's orders?

To sell that $25,000 worth of products, the salesman made 42 visits which required driving a total of 4,000 miles. Special packaging had to be designed. The sales

How the Ratios are Figured

In the ratio tables each group of ratios in each industry carries three sets of figures. The top figure is the upper quartile, the center figure is the median, and the bottom figure is the lower quartile. They are calculated as follows: Year-end financial statements are selected from a sampling of corporations whose tangible net worth, with few exceptions, exceed $35,000. The financial statements are those appearing in the Dun & Bradstreet credit reports on these businesses. Ratio figures are arranged so that the best ratio figure is at the top, the weakest at the bottom. The figure which falls just in the middle of this series becomes the median for that ratio in that line of business. The figure halfway between the median and the highest term of the series is the upper quartile; and the term halfway between the median and the bottom of the series is the lower quartile. The purpose of these interquartile ranges is to show an upper and lower limit area without reflecting the extremes either at the top or the bottom of the series. After the first of the "14 Ratios" has been compiled for a particular industry, the identical process is followed for the remaining 13 ratios in this industry group, and then for remaining industry groups.

$$\text{R.O.I.} = \frac{\text{NET PROFIT (INCOME)}}{\text{TANGIBLE NET WORTH}}$$

CURRENT ASSETS TO CURRENT DEBT
Current Assets are divided by total Current Debt. Current Assets are the sum of cash, notes and accounts receivable (less reserves for bad debt), advances on merchandise, merchandise inventories, and Listed, Federal, State and Municipal securities not in excess of market value. Current Debt is the total of all liabilities falling due within one year. This is one test of solvency.

NET PROFITS ON NET SALES
Obtained by dividing the net earnings of the business, after taxes, by net sales (the dollar volume less returns, allowances, and cash discounts). This important yardstick in measuring profitability should be related to the ratio which follows.

NET PROFITS ON TANGIBLE NET WORTH
Tangible Net Worth is the equity of stockholders in the business, as obtained by subtracting total liabilities from total assets, and then deducting intangibles. The ratio is obtained by dividing Net Profits after taxes by Tangible Net Worth. Tendency is to look increasingly to this ratio as a final criterion of profitability. Generally, a relationship of at least 10 per cent is regarded as a desirable objective for providing dividends plus funds for future growth.

NET PROFITS ON NET WORKING CAPITAL
Net Working Capital represents the excess of Current Assets over Current Debt. This margin represents the cushion available to the business for carrying inventories and receivables, and for financing day-to-day operations. The ratio is obtained by dividing Net Profits, after taxes, by Net Working Capital.

NET SALES TO TANGIBLE NET WORTH
Net Sales are divided by Tangible Net Worth. This gives a measure of relative turnover of invested capital.

NET SALES TO NET WORKING CAPITAL
Net Sales are divided by Net Working Capital. This provides a guide as to the extent the company is turning its working capital and the margin of operating funds.

COLLECTION PERIOD
Annual net sales are divided by 365 days to obtain average daily credit sales and then the average daily credit sales are divided into notes and accounts receivable, including any discounted. This ratio is helpful in analyzing the collectibility of receivables. Many feel the collection period should not exceed the net maturity indicated by selling terms by more than 10 to 15 days. When comparing the collection period of one concern with that of another, allowances should be made for possible variations in selling terms.

NET SALES TO INVENTORY
Dividing annual Net Sales by Merchandise Inventory as carried on the balance sheet. This quotient does not yield an actual physical turnover. It provides a yardstick for comparing stock-to-sales ratios of one concern with another or with those for the industry.

FIXED ASSETS TO TANGIBLE NET WORTH
Fixed Assets are divided by Tangible Net Worth. Fixed Assets represent depreciated book values of building, leasehold improvements, machinery, furniture, fixtures, tools, and other physical equipment, plus land, if any, and valued at cost or appraised market value. Ordinarily, this relationship should not exceed 100 percent for a manufacturer, and 75 percent for a wholesaler or retailer.

CURRENT DEBT TO TANGIBLE NET WORTH
Derived by dividing Current Debt by Tangible Net Worth. Ordinarily, a business begins to pile up trouble when this relationship exceeds 80 percent.

TOTAL DEBT TO TANGIBLE NET WORTH
Obtained by dividing total current plus long term debts by Tangible Net Worth. When this relationship exceeds 100 percent, the equity of creditors in the assets of the business exceeds that of owners.

INVENTORY TO NET WORKING CAPITAL
Merchandise inventory is divided by Net Working Capital. This is an additional measure of inventory balance. Ordinarily, the relationship should not exceed 80 percent.

CURRENT DEBT TO INVENTORY
Dividing the Current Debt by Inventory yields yet another indication of the extent to which the business relies on funds from disposal of unsold inventories to meet its debts.

FUNDED DEBTS TO WORKING CAPITAL
Funded Debts are all long term obligations, as represented by mortgages, bonds, debentures, term loans, serial notes, and other types of liabilities maturing more than one year from statement date. This ratio is obtained by dividing Funded Debt by Net Working Capital. Analysts tend to compare Funded Debts with Net Working Capital in determining whether or not long term debts are in proper proportion. Ordinarily, this relationship should not exceed 100 percent.

Wholesaling

Line of Business (and number of concerns reporting)	Current assets to current debt	Net profits on net sales	Net profits on tangible net worth	Net profits on net working capital	Net sales to tangible net worth	Net sales to net working capital	Collection period	Net sales to inventory	Fixed assets to tangible net worth	Current debt to tangible net worth	Total debt to tangible net worth	Inventory to net working capital	Current debt to inventory	Funded debts to net working capital
	Times	Per cent	Per cent	Per cent	Times	Times	Days	Times	Per cent	Per cent	Per cent	Per cent	Per cent	Per cent
5077	3.29	3.28	16.61	18.65	7.85	11.22	34	9.8	6.6	33.6	48.5	59.0	62.7	15.5
Air Condtg. & Refrigtn.	**2.16**	**1.82**	**9.32**	**13.41**	**5.20**	**5.83**	**49**	**6.5**	**12.4**	**75.5**	**89.3**	**85.5**	**109.0**	**30.6**
Equipt. & Supplies (52)	1.60	1.04	6.25	7.69	3.28	4.23	65	4.9	32.4	149.4	186.8	141.2	155.6	61.8
5013	3.96	3.24	11.85	16.37	5.22	6.59	29	6.7	6.4	30.3	56.3	70.6	45.9	15.7
Automotive Equipment	**2.62**	**1.83**	**7.60**	**8.87**	**3.84**	**4.70**	**36**	**4.9**	**12.1**	**52.2**	**99.0**	**96.2**	**68.3**	**29.6**
(185)	1.92	1.07	3.79	5.14	2.99	3.37	46	3.8	29.4	85.2	154.4	125.9	94.8	51.1
5095	3.29	1.99	13.10	16.95	9.69	13.21	11	13.3	5.4	28.8	68.6	69.9	58.9	8.7
Beer, Wine & Alcoholic	**2.19**	**1.07**	**7.52**	**10.32**	**6.79**	**9.26**	**25**	**9.2**	**18.3**	**65.9**	**156.1**	**94.6**	**91.6**	**27.9**
Beverages (85)	1.54	0.54	4.14	5.44	4.92	6.60	40	6.3	42.0	143.9	234.9	141.6	128.6	49.2
5029	2.63	3.56	12.75	27.30	7.77	15.00	32	17.1	10.9	30.3	58.7	46.9	113.4	14.7
Chemicals & Allied	**1.83**	**1.89**	**8.72**	**13.26**	**5.37**	**7.45**	**43**	**11.3**	**22.2**	**69.3**	**149.8**	**77.2**	**144.7**	**33.8**
Products (53)	1.44	0.46	3.67	6.76	2.34	4.27	53	7.6	49.9	135.8	199.0	113.5	218.2	83.8
5037	3.45	2.17	15.04	20.56	8.07	10.50	22	23.0	2.0	31.0	72.6	38.1	64.1	11.1
Clothing & Accessories,	**2.26**	**1.09**	**7.37**	**10.99**	**6.10**	**6.89**	**37**	**8.7**	**6.2**	**74.9**	**113.2**	**72.2**	**116.4**	**21.1**
Women's & Child.'s (63)	1.59	0.44	3.30	3.99	3.88	4.30	53	6.0	18.9	139.7	157.2	113.2	181.1	33.5
5036	4.10	2.65	11.86	12.39	7.45	7.84	23	9.5	2.4	29.8	73.2	62.0	48.2	9.3
Clothing & Furnishings,	**2.42**	**1.26**	**6.33**	**6.71**	**5.55**	**5.67**	**43**	**6.2**	**6.1**	**63.3**	**90.7**	**82.4**	**80.9**	**25.6**
Men's & Boys' (48)	2.00	0.53	3.49	3.80	3.15	3.66	62	4.9	15.0	102.6	159.3	113.0	98.0	34.9
5081	3.63	2.35	15.70	17.01	7.53	7.68	50	11.1	4.6	46.0	102.9	57.3	64.0	9.3
Commercial Machines	**2.23**	**1.62**	**7.41**	**7.62**	**5.02**	**5.40**	**60**	**6.5**	**9.5**	**71.8**	**133.4**	**79.5**	**109.2**	**28.9**
& Equipment (35)	1.61	0.66	2.23	2.72	3.51	3.68	92	5.6	21.6	136.6	250.9	94.1	198.3	48.0
5045	4.42	1.19	13.17	16.13	16.57	18.92	10	23.5	6.3	23.9	59.3	62.8	38.5	6.7
Confectionery (36)	**2.68**	**0.47**	**7.43**	**8.44**	**9.97**	**11.13**	**17**	**12.8**	**13.0**	**58.7**	**129.1**	**88.1**	**67.9**	**18.1**
	1.56	0.13	1.84	1.85	6.73	7.47	26	8.2	32.0	151.3	250.1	150.8	122.5	46.0

Exhibit 3-3

Wholesaling

Line of Business (and number of concerns reporting)	Current assets to current debt	Net profits on net sales	Net profits on tangible net worth	Net profits on net working capital	Net sales to tangible net worth	Net sales to net working capital	Collection period	Net sales to inventory	Fixed assets to tangible net worth	Current debt to tangible net worth	Total debt to tangible net worth	Inventory to net working capital	Current debt to inventory	Funded debts to net working capital
	Times	Per cent	Per cent	Per cent	Times	Times	Days	Times	Per cent	Per cent	Per cent	Per cent	Per cent	Per cent
5043	3.33	1.72	13.12	30.02	11.32	21.33	16	68.9	17.3	22.6	64.5	22.9	111.0	6.7
Dairy Products (63)	**1.91**	**0.88**	**6.92**	**15.09**	**6.59**	**13.24**	**28**	**29.2**	**46.1**	**57.8**	**111.6**	**38.8**	**228.4**	**37.8**
	1.39	0.38	3.43	6.47	4.45	8.62	34	18.1	60.8	85.4	208.6	76.6	470.3	79.6
5022	2.90	1.56	11.00	12.10	9.97	10.06	29	8.0	6.0	47.4	77.9	77.0	61.5	14.0
Drugs & Druggists'	**2.20**	**1.05**	**6.90**	**7.59**	**5.89**	**7.00**	**38**	**6.6**	**12.3**	**74.6**	**126.6**	**99.4**	**84.6**	**27.4**
Sundries (104)	1.68	0.59	4.15	4.35	4.14	4.68	47	5.6	29.7	140.7	203.8	145.2	122.9	48.1
5064	2.64	1.96	12.00	14.56	8.35	10.17	30	8.1	2.9	54.5	96.2	80.0	76.1	13.2
Electrical Appliances,	**1.81**	**1.25**	**6.95**	**8.61**	**6.23**	**7.65**	**41**	**6.3**	**7.5**	**101.7**	**168.7**	**116.1**	**100.6**	**24.0**
TV & Radio Sets (109)	1.49	0.63	3.60	4.89	4.51	5.06	53	5.1	14.9	163.5	205.1	166.0	148.5	44.6
5063	2.84	2.19	11.75	14.38	8.15	9.02	36	9.8	5.4	44.2	78.6	68.1	70.6	5.8
Electrical Apparatus &	**2.10**	**1.32**	**7.15**	**8.50**	**5.74**	**6.49**	**43**	**7.2**	**12.3**	**76.5**	**122.4**	**86.3**	**105.6**	**19.9**
Equipment (159)	1.73	0.75	3.63	4.40	3.71	4.74	56	5.5	23.9	117.8	178.5	112.1	132.8	35.1
5065	3.86	4.38	15.06	15.38	5.74	6.54	30	5.3	7.2	30.0	76.3	86.0	40.9	12.5
Electronic Parts &	**2.38**	**2.04**	**7.71**	**8.59**	**3.93**	**4.38**	**39**	**4.1**	**11.6**	**62.7**	**104.1**	**106.1**	**61.5**	**28.7**
Equipment (51)	1.96	0.67	3.94	4.10	3.18	3.24	48	3.5	28.3	110.0	149.1	142.9	85.9	42.2
5083	3.78	2.33	13.55	17.35	6.85	7.68	28	8.0	5.6	34.5	72.8	60.0	54.4	10.0
Farm Machinery &	**2.24**	**1.60**	**8.65**	**10.08**	**5.14**	**5.85**	**42**	**5.7**	**12.8**	**63.7**	**117.7**	**97.5**	**82.5**	**22.7**
Equipment (56)	1.67	1.06	3.41	4.29	3.29	3.95	57	4.2	30.1	118.8	170.2	141.0	109.8	46.3
5039	3.36	3.04	13.30	13.38	6.44	7.28	30	8.7	1.0	40.6	69.0	47.8	66.2	4.6
Footwear (64)	**2.15**	**1.68**	**8.17**	**8.26**	**3.95**	**4.87**	**52**	**6.0**	**3.9**	**69.8**	**125.5**	**83.5**	**97.9**	**10.6**
	1.71	0.68	3.23	3.89	2.81	3.52	76	3.8	8.4	114.2	168.8	119.7	149.6	30.3
5048	4.22	1.92	14.94	40.13	13.35	29.57	12	117.8	17.4	17.0	53.4	11.6	89.8	23.6
Fresh Fruits &	**1.83**	**1.32**	**10.17**	**17.06**	**9.79**	**14.17**	**21**	**57.5**	**33.4**	**48.2**	**80.0**	**50.7**	**331.8**	**33.6**
Vegetables (63)	1.30	0.60	4.51	7.48	5.41	9.62	32	21.0	61.4	101.3	146.2	84.4	554.9	69.8
5097	3.62	2.50	13.44	16.77	7.84	7.91	38	11.7	3.5	32.9	105.3	47.9	76.1	10.8
Furniture & Home	**2.37**	**1.65**	**7.88**	**9.12**	**5.07**	**5.81**	**51**	**6.6**	**9.2**	**65.2**	**184.7**	**77.0**	**111.1**	**23.3**
Furnishings (63)	1.52	0.86	3.30	3.71	2.97	3.29	63	4.9	19.1	166.8	256.2	147.4	143.2	38.3
5041	3.61	0.82	10.16	13.52	17.74	23.95	8	15.9	9.1	33.7	84.0	87.9	43.9	18.3
Groceries, General Line	**2.04**	**0.44**	**5.08**	**6.38**	**10.32**	**12.70**	**13**	**10.9**	**20.5**	**78.1**	**144.9**	**121.7**	**73.5**	**38.8**
(215)	1.63	0.16	1.70	2.09	6.81	8.47	20	8.4	55.6	129.8	214.9	164.6	102.6	74.1
5072	4.41	2.25	8.74	11.08	5.87	6.36	34	6.0	5.9	25.1	52.1	72.9	41.7	8.1
Hardware (178)	**2.73**	**1.35**	**4.54**	**5.13**	**3.32**	**4.22**	**44**	**4.6**	**13.9**	**50.0**	**79.2**	**92.5**	**63.6**	**17.9**
	1.99	0.56	2.06	2.39	2.53	2.98	54	3.4	24.1	87.5	156.1	116.5	86.8	41.3
5084	3.24	2.93	14.92	20.47	7.01	9.93	33	11.4	8.6	31.2	61.0	63.2	62.4	14.6
Industrial Machinery &	**2.22**	**1.72**	**8.05**	**10.62**	**5.00**	**6.26**	**43**	**7.3**	**21.6**	**69.8**	**116.9**	**84.6**	**93.6**	**28.1**
Equipment (105)	1.60	0.74	3.87	4.28	3.13	3.78	57	4.7	41.5	110.2	171.6	118.9	164.1	62.2
5098	3.29	2.87	11.90	19.30	8.02	9.95	34	11.2	11.3	31.6	61.9	56.8	62.8	13.4
Lumber & Construction	**2.23**	**1.50**	**7.42**	**9.88**	**4.67**	**6.57**	**47**	**7.0**	**24.3**	**60.9**	**111.8**	**80.7**	**103.7**	**28.3**
Materials (157)	1.59	0.89	4.16	5.21	2.90	3.88	66	4.8	43.1	116.1	178.6	124.9	166.8	44.9
5047	2.83	1.40	15.67	28.76	19.86	29.29	14	64.5	9.0	34.9	76.9	29.2	127.7	29.8
Meats & Meat Products	**1.92**	**0.49**	**7.53**	**12.25**	**14.89**	**21.75**	**19**	**39.6**	**28.9**	**62.5**	**114.1**	**54.4**	**203.0**	**44.8**
(53)	1.69	0.18	2.37	4.10	9.44	15.52	29	24.7	54.7	120.1	189.6	86.7	272.3	88.3
5091	3.55	3.41	10.93	16.05	5.41	7.44	32	8.6	10.0	28.1	63.7	64.9	53.7	10.1
Metals & Minerals (70)	**2.26**	**1.80**	**7.13**	**10.62**	**3.94**	**5.33**	**43**	**5.3**	**24.4**	**56.6**	**88.8**	**96.3**	**86.9**	**26.7**
	1.56	0.90	4.72	7.43	2.90	3.70	51	4.1	44.3	105.3	135.8	142.3	166.9	39.8
5028	5.18	3.53	10.09	14.75	4.03	5.64	32	8.0	8.2	18.0	26.7	45.6	39.0	7.3
Paints & Varnishes (40)	**3.27**	**1.81**	**6.88**	**9.43**	**2.79**	**4.35**	**43**	**6.4**	**13.9**	**28.0**	**37.4**	**73.5**	**69.2**	**14.7**
	2.44	1.30	3.77	4.55	2.24	3.00	53	4.4	26.6	50.5	68.4	90.0	96.6	21.7
5096	3.63	2.09	10.45	13.68	7.99	9.62	29	12.6	5.0	29.2	62.3	55.5	63.2	12.2
Paper & Its Products	**2.45**	**1.17**	**6.35**	**7.88**	**5.29**	**6.41**	**37**	**8.1**	**11.6**	**53.4**	**88.3**	**80.1**	**91.7**	**26.7**
(134)	1.80	0.44	2.98	3.28	3.64	4.75	47	6.0	25.8	96.5	173.6	107.6	125.9	42.8
5092	3.08	3.36	12.29	29.99	7.86	14.24	26	30.4	33.6	23.4	45.8	30.9	99.2	19.1
Petroleum & Petroleum	**1.94**	**2.00**	**8.44**	**15.56**	**4.54**	**8.66**	**34**	**17.8**	**51.4**	**44.8**	**82.0**	**58.3**	**194.5**	**58.3**
Products (86)	1.32	0.69	2.81	5.29	2.74	5.25	53	10.5	81.1	91.3	176.9	96.5	325.3	127.4
5033	2.80	2.59	14.53	16.72	8.81	9.98	34	9.8	1.9	47.2	66.4	61.9	62.3	7.3
Piece Goods (124)	**1.95**	**1.38**	**6.90**	**7.59**	**4.97**	**5.70**	**49**	**6.0**	**4.6**	**103.3**	**112.0**	**99.7**	**103.0**	**22.6**
	1.55	0.79	3.89	3.94	3.37	3.78	70	4.4	10.3	154.0	202.0	136.0	164.2	45.2
5074	4.19	2.98	10.96	13.80	5.33	6.48	36	7.6	6.5	27.5	57.7	63.3	46.4	11.8
Plumbing & Heating	**2.81**	**1.75**	**7.23**	**8.71**	**3.97**	**4.80**	**45**	**5.5**	**13.9**	**47.4**	**89.4**	**79.1**	**73.5**	**21.9**
Eqpt. & Supplies (182)	2.01	0.85	3.81	4.19	2.90	3.42	57	4.4	23.7	76.5	162.3	101.6	102.8	55.6
5044	4.46	1.10	10.68	12.44	15.09	23.46	14	71.7	3.8	18.6	66.1	21.1	84.5	36.1
Poultry & Poultry	**3.12**	**0.57**	**5.04**	**6.85**	**8.92**	**13.94**	**23**	**32.3**	**15.7**	**35.6**	**99.3**	**39.6**	**160.5**	**68.8**
Products (40)	1.70	0.22	1.65	1.75	5.41	7.47	32	16.8	43.2	89.3	159.5	75.5	294.6	93.6
5093	4.39	2.35	9.57	20.00	6.06	10.63	14	23.5	8.6	15.1	41.9	20.1	54.6	27.0
Scrap & Waste	**2.58**	**1.53**	**6.36**	**8.32**	**4.40**	**6.81**	**23**	**12.9**	**30.1**	**37.6**	**61.6**	**45.4**	**115.5**	**38.9**
Materials (50)	1.95	0.60	3.49	4.85	2.64	4.94	37	9.5	56.8	59.3	115.4	83.3	251.9	58.4
5014	3.01	4.27	17.22	26.51	7.28	9.18	31	8.1	8.9	44.1	94.5	69.7	62.2	13.9
Tires & Tubes (37)	**1.85**	**2.39**	**10.82**	**13.62**	**4.27**	**4.80**	**42**	**5.3**	**27.1**	**83.6**	**131.5**	**95.2**	**108.9**	**47.2**
	1.45	1.27	5.96	6.16	2.81	3.83	69	3.8	40.2	168.4	267.9	148.2	169.6	95.2
5094	2.63	0.84	9.72	13.65	22.71	29.09	11	24.5	6.6	48.6	85.5	72.8	66.9	22.0
Tobacco & Its Products	**2.02**	**0.39**	**5.98**	**7.47**	**14.05**	**19.01**	16	18.0	12.4	74.1	149.2	105.9	113.0	37.3
(95)	1.46	0.19	2.92	3.27	9.62	12.46	26	14.0	29.2	140.3	239.2	139.3	153.0	51.6

Exhibit 3-3 (Cont'd.)

manager had to make a personal visit to Soaring Satellite to straighten out a complaint. The Sales Correspondence group wrote 20 letters and sent 20 samples and catalogs. The Order Entry department processed 53 orders calling for 350 line items, and made 15 long-distance phone calls. Your warehouse stored enough inventory to service the account's normal demand. The items were picked and packed and each of the 53 orders had to be shipped at less than truckload rates. In addition, 25 back-orders had to be picked, packed and shipped at premium transportation rates, including 10 air shipments you paid for because of failure to meet promised delivery dates.

Now, how much profit would you say your company made on the Soaring Satellite account?

If yours is a typical company, you can be fairly certain that you have some customers and products that cost you more than they are worth, and others that you make more money on than you realize. Most firms have a good manufacturing product cost breakdown, but work with total average costs when analyzing cost factors beyond the production line. Most firms have a general understanding of the fact that small orders are less profitable than large orders because every order, regardless of size, has to overcome certain fixed costs that are associated with it; i.e., it costs the same to process the paperwork for one unit of a line item as it does to process the paperwork for 100 units of the same line item. But beyond such generalizations, not too many firms have specific data about the relative profitability of different customers and different products.

Frequently companies take it for granted that there is no alternative. The task of precise pricing out of marketing and physical distribution costs seems so formidable that they fall back on the expedient of averaging costs and allocating them to products and customers in direct proportion to volume. If total marketing and physical distribution costs represent 20% of the firm's total revenue, for example, it is assumed that marketing and physical distribution costs on a $1,000 order are $200, and on a $10,000 order are $2,000, regardless of customer, product or other circumstances of the sale.

The fact is, of course, that the costs on one $10,000 order may be $2,000 and on another only $1,000, and under some circumstances the $1,000 order may return more profit to the company than the $10,000 order. The development of new costing techniques, as outlined previously in this chapter, is making it possible for some companies to distinguish between profitable and unprofitable accounts and products, enjoying the distinction of turning losers into winners.

Consider, for example, what the following firms were able to do through the application of refined profitability analysis techniques:

1. A major chemical company designed a profitability report as part of its total order processing system. Cost factors were established for inventory, transportation and warehousing so that the cost for any invoice could be readily identified. Each month the firm published a list of its top 100 invoices. The top 100 also turned out to be the top losers—the least profitable of all the invoices processed. Backtracking, the company found out that out-of-stock conditions on important products in certain key warehouses were causing products to move by illogical routes and methods. Salesmen were placing orders for delivery from

wrong distribution centers. Inventory carrying costs on items carried for a certain few customers resulted in an out-of-pocket loss on a consistent number of orders. These findings formed a basis for enforcing the system and controlling it, and operating it as it was designed to be controlled and operated. The people responsible for sales and those responsible for running the system could readily understand the reasons when they could see what had happened because of their decisions.

2. A distributor of food products found that large orders which moved through his warehouse were unprofitable. When the volume discounts were taken into account, the rehandling, storage and inventory cost more than ate up the margin between cost of goods and selling price. The analysis depicted below showed him that he should bypass the warehouse and make direct shipments:

	Through Warehouse	Direct Shipment
Selling Price	$6,350	$6,350
Cost of Goods	5,840	5,840
Gross Margin	$ 510	$ 510
Distribution Costs:		
Order Processing	4	4
Inventory	12	—
Handling and Storage	90	—
Transportation	330	290
TOTAL DIRECT DISTRIBUTION	$ 436	$ 294
CONTRIBUTION TO OVERHEAD AND PROFIT	$ 74	$ 216

3. A national distributor of electrical products had 95% of his shipment volume in small lots requiring less than full case picking. His profitability analysis showed that 60% of the orders were unprofitable. When analysis revealed that about 20% of the line items accounted for about 90% of the volume, the president complained to the firm's engineers that he thought the cost factors being applied could not be realistic. When the V.P. of Marketing and the engineers reviewed the operation together, it became apparent that the primary problem was in the warehouse, where the amount of walking time for picking and the time and material required for packing small orders made costs so high. After analysis they decided to set up a separate small order picking line, limited to fast moving items and simplified packaging. Elimination of credit checks on orders under $250 removed another item of expense and delay. The high transportation costs on each transaction were overcome by scheduling orders geographically, and by accumulating for consolidated pickup and shipment to specialized distribution carriers. The result of these improvements was that a large segment of the business was converted from a loss to a profit.

4

How to Eliminate Profit Leaks by Improving Your Plant Security

Warehouse security measures represent one of the least recognized areas of profit improvement and operational control in the physical distribution field.

United States' firms lose billions—yes, billions of dollars each year—to both intruder and employee theft. Employees alone account for $5 million per day of merchandise stolen. The insurance industry estimates that 30% of all business failures result from employee dishonesty.

Right now, there is better than a 50-50 chance that significant employee pilferage is taking place in your warehouse.

Just as important, employees—particularly those entering the physical distribution industry work force for the first time—are often corrupted by environments where wholesale thefts occur.

Companies suffer more than direct financial losses from internal theft. Employee productivity has been found to fall proportionately to the degree of misappropriations being experienced; work practices become sloppy, thus compromising the quality of the operation.

Merchandise is frequently damaged deliberately and subsequently purchased by the responsible employees at far less than cost or simply removed from the premises without any payment. Warehouses which support manufacturing operations often find themselves out of stock on items shown by inventory control records to be on hand. Production down-time results as schedules are altered to work around the shortages. This contributes further to profit loss and poor customer service.

Unhappily for the warehouse manager, this rather gloomy picture is far from the exception. We must face up to the fact that this is obscured by condemnations of the reliability of inventory status records, electronic data processing inefficiency and inaccuracy and the so-called poor quality of labor. Indeed, these latter areas can contribute to the overall security problem as well as efficiency of operation.

Solution and control of the security problem, however, lies in a methodical examination of all operational areas within the warehouse and supporting office functions, in the application of appropriate security devices and procedural controls and, most importantly, in a determined, continuing commitment on the part of warehouse management to suppress employee and non-employee theft.

THE SECURITY SURVEY—A PROVEN CONTROL TOOL

The security survey is viewed as covering the entire fabric of the distribution center. It cannot be emphasized too strongly that achievement of good physical security requires proper administrative and procedural controls.

It is viewed as vital, therefore, that the warehouse manager obtain from management a mandate which will permit the taking of a comprehensive security survey. Invariably, such a survey cuts across departmental lines. To avoid interdepartmental friction, it is suggested that a committee be established to conduct the survey. Representatives from Sales, Accounting and Purchasing should be included in the group.

Success in obtaining management's approval for forming the committee and in the subsequent survey, may well depend on the warehouse manager's personality, and how he comes across to others. He should make a great effort to convey the impression that his interest is in the health and profitability of the company, and not in intruding upon the prerogatives of the other executives in the organization.

1. Getting the Survey off the Ground

Having a plan of action in mind when implementing the survey will serve to keep all members of the committee constructively and enthusiastically involved in the project.

As a first step, the Purchasing department should be requested to recommend the inclusion of a security service company representative on the committee. This individual can offer helpful technical advice and cost data which the committee will require in making its final recommendations.

2. Perimeter Survey

An important part of the physical security survey is the careful review of the building (s) and property line perimeter.

An engineering drawing, covering the building and entire facility property, should be obtained. If such a drawing is not available from either the Engineering department or the leaser, a line drawing—however simple—should be prepared. Don't shortcut the preparation of this drawing. Its importance will become evident in its utilization.

Having the facility drawing in hand, and referring to the security checklist in the following section of this chapter, survey the entire grounds and building exterior. Indicate on the drawing all property and building exits and windows. Mark down any violations of good security practice on the drawing. Sequentially number all building doors and windows, and indicate type of aperture and respective security devices in existence; i.e., pedestrian exit with no alarm, fire door with panic bar lock and local alarm, overhead door with sliding bolt, glass display window with foil alarm system, 18-foot-high transom window with no alarm device, etc. Each type of aperture requires individual protection.

3. Internal Survey

Again, using the drawing, examine and record the details of every internal door. Note particularly sensitive, high-value storage areas and their respective security arrangements.

Use the security checklist shown in the following section of this chapter, as you progress from department to department. Document security violations as they are observed. Don't trust your memory.

CASE IN POINT:

International Airline, Main Base Supply Center

The supply center occupies a 300,000-square-foot high cube building and supports several aircraft hangars, maintenance shops, commissary, and aircraft resupply operations. Tractor-trailer trains and automated overhead rail cars are used to effect distribution throughout the complex. A large captive guard force maintains formidable surveillance in all areas.

A. *The Problem:*

In this instance, management provided no definitive dollar amount to shortages that it was experiencing. They requested a comprehensive investigation of employee practices. The results of the investigation, described briefly here, were all too typical of practices found in distribution operations, both in the United States and abroad. They knew they had trouble, but not what or how bad it was.

B. *Employee Malpractices:*

(1) A large percentage of the guard force was found openly stealing liquor, flight bags, food supplies, silverware and other items for their personal use. Even worse, in order to gain support in their efforts to steal, they corrupted other company employees by encouraging them to steal also.

(2) A supervisor boasted to an operative of stealing, among other things, aircraft carpeting. What was the carpeting used for and how did he manage to remove it from the premises? Well, the supervisor owned a commercial dog kennel, and he used the carpeting to cover the floors of his dog cages! Rather elegant—but at company expense, why not? The carpeting was cut to the dimensions of the cages on company property and company time, and removed piece by piece.

(3) Another supervisor was so totally disdainful of his responsibilities and of

company merchandise, that he ordered cleaning personnel to sweep all aircraft parts having fallen to the floor during the order picking process into the trash each day.

(4) Storemen and supervisors were found gambling for money on company time. Considerable productive effort was thereby lost. The corruptive influence of this gambling on the entire work force was beyond measure.

(5) Employee malingering was widespread. Some typical observations:
- many employees ceasing work at 3:00 p.m.;
- storemen found reading pilfered magazines at odd times of the regular workday;
- fork-lift operators taking hour-and-a-half smoke breaks;
- day-shift shipping supervisors complaining that the night shipping crew performed virtually no work.

(6) Fork-lift truckmen and personnel-transporters were observed racing their vehicles and damaging aircraft parts. In a typical misjudgment, while making a turn into an aisle, an engine cowl was severely damaged; the cowl was valued at $48,000.

(7) Employees used requisition forms to obtain supplies for their personal use. All requisitions, even for the most costly instruments, were filled from stock without question. Often the requisitioner would merely send the form to the Stores department through the vacuum tube delivery system. The item would then be sent to the requesting party via the automated overhead rail transport system. In one instance, an undercover operative who was working in the storeroom found that a requisition carried *his* employee name and number. He showed the requisition to his supervisor who merely told him to send the parts through.

(8) Overtime payments were found to be completely out of control. Employees bragged about being paid for overtime but not performing any work at all.

(9) Employees punched each other's time cards in order to facilitate their arriving at work late or leaving early.

In summary, the evidence indicated that a serious breakdown of management control over the workforce had occurred. Symptomatic of the problem itself was the fact that for political and other reasons, management felt constrained from taking any effective action.

The incidents illustrated above are, in their statement, suggestive of the remedial action required.

4. Systems and Procedures

This is the least straightforward and, therefore, the most difficult part of the survey. It calls for detail work associated with creative thinking.

As an important aid in conducting this portion of the survey, the preparation of procedural flow charts is strongly recommended.

An illustration of a procedural flow chart is shown in Exhibit 4-1. Notice that it covers the interaction of several departments' activities in controlling a particular activity.

The flow chart example shown covers the operations of a large distributor of restaurant supplies in the United States. Note the effort made to achieve security checks and balances. This enhances control and, therefore, security.

For example, under credit memo flow (see Exhibit 4-1), observe that the Sales department is the authority which evaluates and then issues the return authorization. The Shipping department then arranges for picking the material up and evaluating its condition. Inventory Control then records the transaction into the perpetual inventory records. The Accounting department subsequently reviews the verifying documentation, and also maintains a customer return file—a good check on the frequency of customer returns, operational conditions which may be resulting in returns and salesmen's practices.

CASE IN POINT:

Air Conditioning Manufacturing and Distribution Center

This company employs 1,500 people on two shifts. Annual sales—over $200 million.

A. *The Problem:*

Auditors reported $500,000 inventory loss over three years. All efforts to find the source of shrinkage failed.

B. *How Theft Was Accomplished:*

(1) It was determined that Shipping and Maintenance department personnel were working in collusion with a junk dealer. Shipping department personnel would divert good units through the Maintenance department, which regularly disposed of scrap metal from the manufacturing operation. The junk dealer was, of course, the high bidder on scrap metal in all categories since he could count on receiving valuable commercial and residential air conditioners for 10 cents a pound—the rate paid for scrap iron.

(2) Aside from the misappropriations described above, investigation also uncovered another surprising source of loss to the company. One of several undercover operatives assigned to the investigation was invited into a small ring of warehouse employees who were working in collusion with the Purchasing department manager. The group took advantage of a generous company policy which permitted the sale of damaged units to employees at cost or below, depending on the extent of the damage. They would deliberately create minor, superficial damages to cabinets, thus qualifying the units for special sale to themselves. The Purchasing manager was the authority entrusted by management to establish the charge to employees for damaged units. Quite understandably, his cohorts obtained excellent deals. Our creative Purchasing executive arranged for the disposal of the units through friends in retail outlets, and he shared the profits with members of the ring.

C. *Solution:*

(1) It was recommended that perpetual inventory records and physical counting procedures be improved. As part of an effort to recognize shortages at the earliest possible date, a cyclical inventory taking procedure was introduced where random, spot inventory counts were effected throughout the year. Prior practice called for a single, annual physical count.

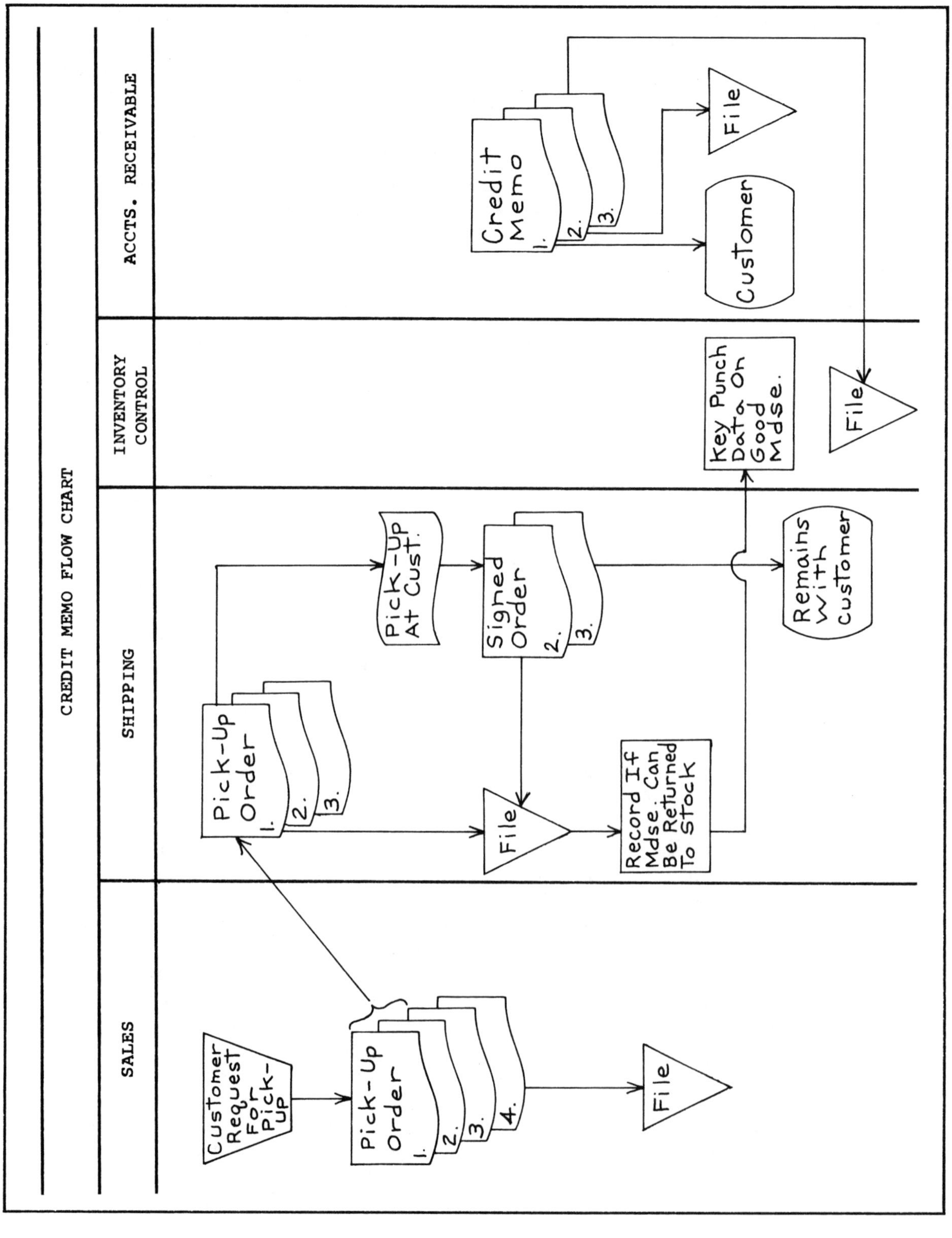
CREDIT MEMO FLOW CHART
SALES
SHIPPING
INVENTORY CONTROL
ACCTS. RECEIVABLE
Customer Request For Pick-Up
Pick-Up Order
1.
2.
3.
4.
File
Pick-Up Order
1.
2.
3.
File
Record If Mdse. Can Be Returned To Stock
Pick-Up At Cust.
Signed Order
2.
3.
Remains with customer
Key Punch Data On Good Mdse.
File
Credit Memo
1.
2.
3.
Customer
File

Exhibit 4-1

(2) A procedure was established whereby a detailed manifest was required for all scrap removal from the premises. Gate guards were required to examine the loads and verify them against the manifest.

Since the guard force was supplied by an outside agency (an advantage in our opinion), it was suggested that management insist that the guards be periodically rotated out of the company.

(3) The case of the erring Purchasing manager was particularly depressing to management since this rather high-level individual had considerable latitude in his actions. Rather than undermine employee relations by ending the employee purchase plan, the following action was taken:

a. Employee purchases were limited to one per person per year.
b. An employee sales record was maintained and audited by the Accounting department.
c. A price list was established by unit, thus eliminating favoritism.
d. The Quality Control department was called upon to evaluate all incidents of product damage.

5. Shift Security

If yours is a multi-shift operation, it will be absolutely necessary that an internal and external physical security survey be made for each operating and non-operating shift. Experience indicates that even good first shift security practices tend to fall apart during second and third shift operations.

CASE IN POINT:

Meat Packing House

This firm employs 500 people on a three-shift basis. It operates 150 wholesale delivery vehicles which cover a six-state region.

A. *The Problem:*

The firm's management reported $375,000 annual loss in unexplained shrinkage. The shrinkage was felt to be occurring between accumulation of orders on the shipping dock and point of delivery. Drivers reported shortages while Shipping department personnel provided documentation to support their position that all merchandise was loaded onto the company's vehicles.

B. *Operating Procedures:*

(1) Delivery trucks were loaded on first and second shifts and moved into the parking area where they were connected to the coolant supply.
(2) Drivers arrived during the third shift. Load manifests were presented to the guard by departing drivers at the only exit/entry gate.
(3) The guard station at the parking lot gate was manned 24 hours, seven days a week.

C. *How Theft Was Accomplished:*

A small group of drivers made it a point to arrive early for the third shift. They would then proceed to unload merchandise from other vehicles onto their own trucks. The guards could not possibly reconcile actual contents against manifests since loads were in 20-foot to 40-foot trailers. They were only serving as effective deterrents to total vehicle theft by intruders.

Drivers sold the merchandise through their own personal customers who, ironically, were served first since their items were in the rear of the trailer.

D. *Solution:*

Shipping department supervisors were required to apply pre-numbered, tamperproof seals to truck doors immediately after loading. The seal number was entered on the respective manifests with copies sent to the guard station. Guards were required to compare seal numbers against those shown on the manifests.

A shortage record was maintained by the driver and also by loaders to highlight habitual problem personnel.

Truck loaders were rotated through other jobs in the warehouse to reduce the probability of collusive arrangements. Also, in the latter connection, loaders were rotated through the various delivery route stations.

SECURITY CHECKLIST

The following checklist represents the most comprehensive coverage of a very complex subject. It will be noted that warehouse security is considered to be dependent on far more than physical devices and procedures.

We feel strongly that security controls are not merely a matter of locking doors and examining packages. In the final analysis, control is a human relations problem. Management's attitude and actions can affect the problem. If management is indifferent to employee morale and practices, misappropriations are thereby encouraged.

Uniform and firm application of the rules suggested in the following checklist will enhance employee respect for their management and reduce the temptation to steal.

The great majority of company employees start out honest. One of management's fundamental responsibilities is to help keep them that way through the application of "loss prevention" techniques.

I. INTERNAL SECURITY

Internal security should be considered as serving as a deterrent to employee theft as well as to detect the presence of intruders.

1. Employees

(a) Pre-employment checks should be conducted. This should include personal calls to prior employers. Important intelligence often results from discussions rather than from written communication. Because of possible legal complications, many firms hesitate to write derogatory background statements. A credit check is helpful and local police often assist in identifying prior offenders.

(b) Identification badges should be required and worn conspicuously.

(c) Employee badges and passes carried by non-employees should be inspected by guard force or supervisory personnel.

(d) Where logical, employees should be restricted to areas where they need to be in order to conduct their business. Storage areas, shipping and receiving, returns handling and other sensitive areas should be strictly off-limits to unauthorized personnel.

(e) Employees should be encouraged to challenge strangers. Protecting company assets should be considered everyone's job.

(f) Requiring the filling in of a Fidelity Bonding Form as part of the pre-employment procedure serves as a deterrent to malpractice. Experience has shown that 20% of applicants leave the employment office without offering any explanation when they are required to complete a bonding form. The bonding form should be used even where the company chooses not to incur the expense of actually bonding employees.

(g) Passes should be provided for any parcel removal. Authorized pass signatures should be established with facsimiles provided to the guard force. All passes should be sequentially numbered and the respective pass numbers registered by the issuing supervisor, and also by the reviewing guard.

(h) Pocketless uniforms should be issued at company expense in firms where merchandise is costly and small or otherwise desirable. Drug warehouses would represent a typical application.

(i) Employees should be provided with lockable cabinets to secure their personal valuables but these cabinets should be subject to spot inspections.

(j) The following activities by employees should alert managers to possible security risks:
   (1) The borrowing of large sums of money.
   (2) Gambling or association with known gamblers, bookies, number writers or other undesirables.
   (3) Excessive employee drinking.
   (4) Little deals with shippers and receivers. Often these little deals grow into big collusive theft arrangements.
   (5) An employee living beyond his income.
   (6) Excessive attachments or garnishments.
   (7) Regular shortages on a particular delivery driver's route.

2. Physical Security

(a) An economic analysis should be conducted to help evaluate the tradeoffs between guard force and security devices. In larger firms, a combination of guard presence with accessory systems to improve coverage and effectiveness is usually found to be most desirable. Smaller enterprises usually rely on alarm systems with service company monitoring.

(b) All outside access doors should be alarmed and wired to a central monitoring office.

(c) Any internal door providing access to sensitive areas during non-working periods should be alarmed and wired to a central monitor.

(d) Remote TV cameras can minimize total security cost. Through their use, a centralized guard station can monitor many sensitive areas within the warehouse; i.e., shipping and receiving docks, storage areas, entry doors. Their mere presence serves as a deterrence to misappropriation by employees.

(e) Waste removal should receive intensive scrutiny. Compaction is favored today as a "clean" method of removing wastes while simultaneously inhibiting any employee desiring to secrete merchandise for eventual removal from the premises.

(f) Sanitary and other facilities for visiting truckers should be located away from storage and dock areas.

(g) All truck dock doors should be closed and locked during lunch and break periods and after business hours. A supervisory presence on the truck dock is most important during operating hours. Supervisory lunch periods should be rotated to facilitate this presence.

(h) Employee entrance/exit should be limited to a single access point.

(i) Delivery trucks filled for next day routing should be locked and sealed, with a tamper-proof, numbered seal. The seal numbers should be entered on each vehicle's manifest and the number checked by the gate guard. Some firms having extremely valuable merchandise have modified their vehicles so that they can be easily connected to a central alarm system. Any tampering with truck doors raises a central station and local alarm.

(j) Make strenuous efforts to avoid damages to packages. Open cartons are easily pilfered. If damage does occur, reseal immediately.

(k) Listed below are a few indicators which suggest internal dishonesty:
   (1) Broken cases with some of the contents missing.
   (2) Merchandise found in unusual locations, such as near an exit, in a restroom or in a waste receptacle.
   (3) A key that does not fit the lock for which it was intended, thus indicating that the lock was switched.
   (4) Electronic security devices frequently in need of repair or inoperable.

(l) Review your physical security methods and systems annually to determine if coverage is still adequate. Efforts to cut down cost by concentrating on a few major entry points of the building may result in a break-in at those points that have been neglected. Technological developments in recent years have resulted in many inexpensive, local alarm devices which can be used to supplement the main security system.

(m) Provide security for all windows through wiring, screening and/or barring.

We are concerned with more than unauthorized entry. Employees have been known to throw merchandise out.

(n) Shrink-wrap broken case lot shipments. This provides the customer with better security at his end, thus reducing the possibility of his reporting shortages and requesting credit.

(o) Provide warehouse supervisors with transportation to facilitate their supervisory efforts. Personnel carriers are relatively inexpensive. Supervisory presence is the best insurance against pilferage.

3. Outside Service Personnel

(a) Background checks should be conducted by your firm or by the contractor where such persons regularly service the premises.

(b) Management should require that service personnel be bonded.

(c) Supervision of service personnel is vital. They should be prohibited from sensitive areas. If they must enter storage or other pilferage-sensitive locations, it should be under supervision.

(d) Upon departing the premises, all service personnel should be checked. Any parcel pass or lunch box should be examined. Parcels covered by passes should be examined and contents compared against pass description.

4. Vendors

Vendors often serve us as sources of technical and market information. In this capacity, they are often invited to review many areas of warehouse operations. We would in no way wish to inhibit their usefulness. Certain fundamental rules should, however, be enforced:

(a) They should be escorted while in the warehouse.

(b) They should be identified as visitors through distinctive badge identification.

(c) They should be logged in and out of the premises. Briefcase and parcel checks should be effected on departure.

5. Inventory Control

Recognition of a shortage at the earliest possible moment is in itself, an important deterrent to employee malpractice. Inventory control records and procedures play an important role in this connection. A daily perpetual record should be maintained. Further, to test record accuracy, a cyclical inventory checking procedure should be followed. This procedure conforms with accepted accounting practice. It replaces annual inventory taking methods with periodic counts of all items on a random basis over the entire year. Errors or discrepancies tend to be recognized earlier, thus alerting security personnel to problem areas.

6. Controlled Error Tests

Introducing deliberate errors into shipping and receiving documentation can help you evaluate supervisory and rank-and-file attention to detail. Employees should be told that such tests will be conducted. The purpose here is to encourage alertness, not to trap people.

7. *Guard Force*

(a) Screen guards carefully. Use a Fidelity Bond.

(b) Set guard force wages at a level which will not encourage their participation in pilfering.

(c) If an outside guard force is used, require guard rotation to avoid familiarization with employees.

(d) Require guards to take polygraph tests.

(e) Alternate guard routes, timing.

8. *Protection of High-Value Items*

Small and valuable items should be segregated in one central location that is well lighted and always in view of supervisory personnel. Scattering such items throughout the warehouse increases the probability of pilferage.

9. *Shipping*

(a) *Responsibility*—Where economically practicable, supervisory responsibility for shipping should be separated from receiving and warehousing. Sufficient evidence has been accumulated to demonstrate the impropriety of having one individual responsible for both shipping and receiving. The temptation for collusive malpractice is too great.

Where the type of business dictates that one individual have responsibility for both shipping and receiving, another employee should maintain a perpetual inventory record. Random spot checks on quantities loaded into delivery vehicles against shipping manifests should be made.

(b) *Reduction of Driver Shortages*—Itemized shipping documents should be signed by both the shipping clerk and the drivers at loading time.

To reduce the possibility of collusion between drivers and loaders, the latter personnel should be rotated to other jobs within the warehouse.

Undeliverable merchandise should be returned by the driver to the shipping supervisor immediately upon his return to the warehouse. The merchandise should then be re-entered into inventory.

(c) Only company delivery or customers' pickup vehicles should be permitted in the shipping dock area.

(d) Only merchandise covered by shipping documentation should be permitted on the shipping platform.

(e) Responsibility for customer order entry should be separate from Shipping department responsibility.

(f) Shipping documentation should be serially numbered with separate copies sent to Accounts Receivable and to Inventory Control.

(g) All quantities of damaged or otherwise unsaleable merchandise should be verified by someone other than warehouse personnel. Proper reductions from inventory should then be effected.

(h) Maintain records of all claims by customers on short shipments. Cross-reference this record to respective drivers. Also, have packers include a slip with their name or number as a further feedback control on short shipments.

10. Receiving

(a) Only suppliers or company vehicles should be permitted in the Receiving area. Employee parking should be in a location remote from Receiving.

(b) Received goods should be removed from the platform and placed into stock immediately following count verifications.

(c) Pre-numbered forms should be used in detailing verified counts. "Blind" receiving documents are preferable. These show item descriptions as prepared by Purchasing but quantities are omitted from the Receiving copy. This forces a count by Receiving personnel. The Receiving copy should be reconciled against the Purchasing copy by the Accounts Payable department.

(d) Receiving department personnel should not be permitted to work in the Purchasing department.

(e) Returned goods should be verified by Purchasing or Accounting department personnel.

(f) Unissued warehouse receipts should be checked regularly by a representative of the Controller's office to determine that none are missing. The Inventory Control section should maintain a numerical listing of receiving reports to assist in this reconciliation.

(g) Place all returned goods which require repacking or refurbishing into a high security room. Maintain separate perpetual inventory records of merchandise in this area.

II. EXTERNAL SECURITY

External security arrangements should achieve controlled access to the warehouse proper and to other sensitive areas within the facility perimeter; i.e., vehicle parking locations. Good security need not inhibit normal business activities.

1. Illumination

Good lighting of exterior grounds is a strong deterrent to the "casual" thief and to malicious mischief. Coupled with central station, remote T.V. viewers, it can reduce guard force requirements without sacrificing coverage.

2. Outside Fencing and Alarm Systems Functioning

Too often costly security systems are compromised through neglect of maintenance. Periodic checks of these systems are strongly recommended. If guards are used, their normal routine should require listing any break in perimeter security. This would include: cuts in fences, padlocks left unsecured, window screening unlocked and so on.

3. *Are Guard Patrols Adequate?*

A major failing of many security systems is the lack of adequate coverage during the early morning hours and weekend periods.

Since many insurance companies require a "fire watchman" when the premises are not otherwise occupied, it is recommended that this individual serve a dual role of security guard.

4. *Are Contents of Exiting Cars and Trucks Checked?*

Employees' and visitors' vehicles should be inspected on a random, spot-check basis. Trunks should be opened. Shipping and Receiving department supervisors should periodically examine a departing vehicle's load against its manifest. The latter inhibits collusion between employees and truckers.

5. *Key Control*

Keys to all doors and gates should, in turn, be kept under lock and key in a supervisor's office, not left on a peg board in the shipping area or other unsecured location.

Only a minimum number of executives and members of the security force should have access to these keys.

6. *Magnetic Locks and Card Controls*

Magnetic cards, in lieu of conventional locking devices, are used where one wishes to permit normal employee traffic while excluding unauthorized individuals passage. A typical application is employee parking lot gate control during working hours. This device has proven to be a strong deterrent against pilferage from cars as well as muggings during night shifts.

These devices, which are impossible to duplicate, are also used where it is desirable to limit access to sensitive areas to a few key personnel. High-value, segregated storage areas lend themselves to this approach.

7. *Pedestrian Exit and Entry Control*

Many older warehouses located in urban areas with high crime rates, have a particular problem with a multiplicity of street exits. Guards stationed at each exit would, of course, solve the problem. However the cost would be too high.

A very effective device which has helped solve this security problem is the "Dead-Bolt-Lock-with-Pressure Panic Bar." This device prevents entry through a pedestrian door but permits personnel to exit in case of emergency.

This unit can be equipped with a local, battery-powered alarm wired to a central guard station alarm panel, and can also be equipped with a recorder to indicate date, time and number of openings. A special, non-reproducible key is supplied to enable security personnel to test and shut off the local alarm or to exit without sounding the alarm.

The device also has application in modern warehouses where insurance companies require emergency personnel exits in addition to those required for normal business operations.

5

Examining Warehouse Operations for Cost Reduction Possibilities

This chapter presents four operational areas where important dollar savings and improved customer service may be obtained:

- How to conduct a systematic survey of existing operations and improve methods and control.
- Examples of proven warehouse control reports used to reduce direct labor expense.
- Ways to reduce customer order cycle time.
- A freight expense analysis system which can significantly reduce transportation costs.

A. SYSTEMATIC AND PERIODIC OPERATIONAL REVIEW HELPS MAINTAIN DISTRIBUTION SYSTEM EFFICIENCY

One of the most difficult tasks facing the physical distribution manager is critically evaluating his own operations. But periodic, in-depth, operational reviews are necessary—once each three to five years is a good interval. Such reviews are desirable for several reasons:

- Personnel turnover can result in new and poor working habits.
- Other departments modify their procedures and these can unknowingly affect distribution operations. For example, changes in purchasing practices can result in increased rail car unloadings and demurrage payments. A change in customer payment terms by the Financial department could result in a significant increase in small shipments and the need to maintain larger inventories.
- New materials handling equipment developments may be applied to your operations with substantial savings.

CASE IN POINT: During an operational survey, a soft-goods distributor found that his volume of business had increased to the point where an overhead conveyor system for transporting garments could save $55,000 per year in handling costs. Aisle congestion was eliminated and product damage was reduced also.

- Good security practice calls for periodic review as a deterrent to malpractices which can occur in even the best managed operation.
- Finally, if you are just taking charge of a new operation either through promotion or change in company affiliation, an operational study should be the first order of business. Conducting an in-depth review is the fastest method of getting to know the strengths and weaknesses of your subordinates and your facility—it puts you "on top" immediately.

B. THE TYPICAL OPERATIONAL SURVEY

The discussion below covers an increasingly common situation in business today. A cosmetics distributor was acquired by a conglomerate and a new executive staff brought in to manage the company. The first order of business was to conduct an operational survey to determine how efficiency and profitability could be improved. Important areas of analysis and results of that survey are discussed below. The approach and benefits may be applied to any distribution operation.

1. Inventory Control

Just prior to the acquisition being finalized, the controller of the company decided to replace a manual, perpetual inventory card system with a computerized system. Rather than having the old system run parallel with the new one until the "bugs" could be worked out of it, the decision was made to scrap the manual system. Unfortunately, this proved to be a premature and costly decision. At the time of acquisition, no reliable inventory status information covering the company's 200 finished goods items was available.

The manual inventory record card shown in Exhibit 5-1 was recommended as a short-term solution to perpetual inventory record keeping and inventory planning. A review of the computerized system showed that a complete reprogramming job, requiring several months' work, was necessary. Further, the system concept was not adequate for inventory *management* purposes. It is important to note that the value of a computerized approach to inventory management is not merely in the generation of a perpetual

FINISHED GOODS

INVENTORY CONTROL CARD - MANUAL SYSTEM

STOCK NUMBER: 0100

ITEM DESCRIPTION: AIR MOISTURIZER

ORDER POINT: 50
ORDER QNTY : 1500
COST : $90.00/M

DATE	DOCUMENT NUMBER	OPENING BALANCE	(+) RECEIPTS	(-) ISSUES	=ON HAND	PLANNING					
						(+) ON ORDER		(+) HELD AT VENDOR		=AVAILABLE	AT ORDER POINT
						P.O. NO.	QNTY.	VENDOR CODE NO.	QNTY.		

Exhibit 5-1

inventory record. Rather, a computerized application can be economically justified more readily when the basic inventory status information is used for such purposes as: computing usage trends from historical and forecast data; determining economic lot sizes for Purchasing; totaling usage by item for given time periods and comparing against prior periods; calculating turnover rates; calculating weeks of inventory on hand by item, and providing other information on which to base inventory, purchasing and sales management decisions.

The lack of a reliable inventory data base and the substantial effort which would have been required to construct a computerized inventory management system strongly suggested that a manual system be implemented first. The following steps were recommended to establish and properly maintain a Manual Perpetual Inventory Record:

(a) Take a new physical inventory.

(b) Establish management responsibility for the inventory management system. It was recommended that inventory management be assigned to the Purchasing manager.

(c) Indoctrinate warehouse personnel in the need for reporting accuracy.

(d) Introduce the inventory system to Sales management, stressing the need to work within the system and not bypass formal procedure.

(e) Request periodic, random spot physical inventories at the warehouse to verify inventory records. This procedure also serves to eliminate the need for the traditional, annual physical inventory taking.

(f) Establish order points and order quantities through cooperative effort between Sales and Purchasing management.

RESULT: Within 12 months after implementing the Manual Inventory System, inventories were reduced by 28% for a dollar savings of $300,000. Just as importantly, customer service was substantially improved.

2. Component Parts Assembly

Fifteen people were employed to assemble finished goods and components (pouches, containers, packaging materials, etc.) into special "deals" and promotions. In this way, new finished goods were created from internal stock. An unnumbered production report was used to document those items used and record the resulting new additions to inventory.

Two recommendations were made to improve the accuracy and reliability of data generated in the Assembly department.

(a) As a matter of good inventory control and security, production reports should be sequentially numbered and a log of the issued and subsequently received numbers maintained by the inventory control clerk. In this way, any misplaced or lost report will be recognized and the warehouse copy referred to. Proper record keeping here may not appear to be too important at first glance.

But poor record keeping often results in destroying the reliability of inventory management systems. And when this happens, shrinkage cannot be detected—it is, rather, explained away as inaccurate data.

(b) The Production Report was also used to process customer returns. To facilitate analysis of returns and regular production work, it is always better to use separate, pre-numbered forms.

RESULT: Documented savings in reducing annual inventory shrinkage by 5%, or $57,000 per year, was achieved. As has been observed in numerous other companies, good record keeping is in itself an important deterrent to misappropriation.

3. Customer Returns

The survey showed that very poor control was being exercised over returned goods from customers. The company was very lax in processing returns which amounted to $200,000 annually. Individual items were valuable. The wholesale value of an ounce of perfume, for example, was $25, and alligator handbags retailing at $185 were also in the product line. As expensive as the merchandise was, the great volume of returns overwhelmed the capacity of the warehouse. Rather than properly document, segregate and process returns, they were hurriedly received and "dumped" wherever space was available. Many customers complained of long delays in receiving credit and refused to pay current bills until their accounts were settled. Pressure was then brought to bear on the warehouse manager who merely accelerated paperwork approval—the respective merchandise remained in scattered piles. The warehouse manager had some revealing complaints of his own:

- He maintained that the great bulk of returns were received without "Return Authorization Forms" as required by company policy. The effect of this was that many customers habitually bought large quantities to obtain price breaks, and then returned the bulk of the order without suffering a price penalty.
- He also pointed out that a substantial number of customers ordered promotional items which were accompanied by "free gifts." The promotions were being returned in large quantities *minus* the free gifts. The total annual value of these gifts amounted to $45,000.
- No log or other record was maintained to record how expeditiously the warehouse processed returns.

Recommendations and Results:

—Authority to issue return authorizations was fixed within the Sales department. A central file covering returns was maintained in the Accounting department. Any unauthorized return was referred back to the Sales department for review before credit was applied to the customer's account.

—Upon review, management decided to eliminate the free gift with purchase.

Other sales incentives were used which did not lend themselves to such heavy abuse.

—All returns were promptly processed and customer accounts credited. Good merchandise was returned to regular stock. Damaged goods were pulverized and scrapped.

—Sequentially pre-numbered forms were designed to properly document returned goods transactions.

—A fenced, high security area was established to segregate and process returns.

—Improved controls and processing of returned goods as described above resulted in the salvage of more units for resale; reduced losses through customer abuses and employee misappropriation and facilitated better customer service.

4. Customer Order Entry

Customer orders were generated by the following sources:

- Written orders mailed by customers.
- Field salesmen and regional sales offices.
- Phone orders from any of the above sources.

All billing was accomplished at a headquarters office, remote from the warehouse. The company pre-paid all shipments over $350 and sent pre-billed invoices to the warehouse for order selection. Back-orders required manual adjustment to invoices prior to shipment. The latter adjustments introduced an additional error factor and created an unbusinesslike document.

No record of back-order frequency (customer service levels) was maintained.

Recommendation:

A post-billing procedure was recommended for the following reasons:

1. To reduce the incidence of errors resulting from manually adjusting invoice totals when back-orders occurred.
2. Customers did not require an invoice with shipment.
3. Manually adjusted invoices present an unbusinesslike appearance.
4. Pre-billing is technically more cumbersome to operate than post-billing when headquarters operations are remote from the warehouse. This is due to the accounts receivable adjustments that are required when back-orders occur.

Further, the suggestion was made that until such time as post-billing was introduced, the Accounting department verify all manual entries to invoices. Finally, it was recommended that a "Daily Customer Service Record" be maintained by the Inventory

Control section for periodic review by Sales and Materials management. The details of several approaches to calculating and displaying back-order information are discussed in another chapter of this book.

C. INFORMATION REQUIRED TO ESTABLISH DESIGN OF ORDER FILL SYSTEMS

Product:

1. A complete list of commodities to be handled, with dimensions and weights of each and a description of any special characteristics of the commodities, such as fragile goods, frozen items, soft cases.
2. Frequency of shipment of each item; i.e., the hourly, daily, weekly, monthly and yearly requirements and variations.

Orders:

1. A cross-section of the hourly, daily, weekly, monthly and yearly requirements and variations.
2. Minimum, average and maximum figures for order size and items per order.
3. Method of preparing order; whether on multi-copy paper forms, machine-punched cards or tape.

Physical Limitations and Building Information:

1. Description of the building and space available for the system.
2. Floor plan of the area where the order fill system is to be located. Dimension and locate columns. Indicate on floor plan the maximum available space in which storage lines can be located and note any aisles desired in this area.
3. Indicate on plan the area where product is stored before being placed in order fill system, and shipping area where orders are to be sent after makeup.
4. Availability of electric current and air supply with characteristics.

Loading of Storage Lines:

1. Are storage lines to be loaded manually or automatically?
2. Are commodities brought to storage lines on pallets?
3. Indicate the total number of items per pallet for each commodity.
4. If commodities arrive on conveyors, do they come in groups of like items or at random?

Order Filling Operation:

1. Are part case lots involved?
2. Are all of one commodity to be in a group when coming off collect system?
3. Is a complete order count required?

4. Is order checking to be eliminated?
5. Is order separation required?
6. How many hours per working day desirable?
7. Is first-in, first-out control of primary importance?

Shipping Operation:

1. Designate the type of shipping system desired.
2. Are orders stacked on pallets, delivered direct to truck or car or stored on conveyors for future shipment?
3. Is destination marking required?
4. Do shipping operations and order filling operations coincide as to time?
5. If orders are delivered directly to trucks, determine the number of trucks to be loaded at a time.
6. Number of cases and orders per truck.

Present Operations:

1. Number of lift truck operators?
2. Number of order pickers?
3. Number of stockers?
4. Number of checkers?
5. Number of shippers or truck loaders?
6. Number of supervisors, dispatchers, etc.?
7. Number of labelers?
8. Rates of each of the above in terms of cases or pallet loads handled per net hour worked.
9. Actual working hours in an eight-hour shift?
10. What type of order form is presently used?
11. Describe present stacking, picking and shipping operations.

Future Operations:

1. Indicate anticipated expansion in two years, five years.
2. Is proposed system to have capacity for immediate expansion or are provisions to be made for additional equipment later?
3. Indicate any future changes such as change in order form, case sizes, order size and pallet size which might materially affect operation of system.
4. Indicate frequency of addition of new products. Do these replace older products?

D. TRAFFIC COST ANALYSIS

Consolidation of shipments to customers represents an area of great potential savings for many companies. The E.D.P. exhibits presented in this section are designed to assist you in analyzing your traffic costs to determine if there is an opportunity for consolidation. These computer print-outs cost thousands of dollars to develop. Your own Systems department will find no difficulty in applying the basic formats illustrated to your own operation. The two major factors critical to embarking on a freight consolidation program are the number of shipments made to individual geographic areas within a narrow time span and the respective weights of those shipments. Exhibit 5-2 shows how to break down your shipment data to evaluate the potential savings which consolidation may bring.

CASE IN POINT: The analysis of freight costs as shown in Exhibit 5-2 is from an actual study performed by a tobacco products distributor whose total annual freightout bill amounted to $2.5 million. The analysis demonstrated that an annual savings of $500,000 could be achieved by *expanding an existing consolidation program* within the company. Expansion of the consolidation program required two actions:

(A) The Sales department agreed to "sell" certain customers on the desirability of "consolidating" their orders to the company. This was designed to reduce the number of shipments per week to individual customers.

(B) Certain order entry functions such as credit checking, application of customer account numbers and data verification were to be computerized to save time in the order entry cycle. Rather than use this time savings to get orders out more quickly, the time (two days) was used to consolidate shipments. A survey of the company's customers indicated that shortening the order cycle would not result in increased sales.

Listed below is a brief explanation of some of the less obvious column headings shown in Exhibit 5-2:

"ST" — State
"ZIP" — Three-digit zip code representing customer location.
"3 OR LESS LINE" — The number of customer orders which contained three or less line items.
"TOT" — Total orders shipped to respective customers in five-month period.
"SAME DAY" — The number of times two separate customer orders were shipped on the same day.
"PRIOR DAY" — The number of times orders were shipped to the same customer on two consecutive days.
"WEEK" — The number of times two or more orders were shipped to the same customer within one week.

The Shipments Report shown in Exhibit 5-2 is useful in analyzing individual customer shipments. Totals are also shown by zip code and demand areas, which may be considered as destination points for consolidation purposes.

V-000 ORIGIN REPORT COVERING SHIPMENTS FROM 08/20/ THRU 01/20/ 1000107-02

CUSTOMER NUMBER	NAME	ORIGIN	DEST ST	ZIP	NUMBER OF SHIPMENTS: 3 OR LESS LINE	TOT	SAME DAY	EXTRA PRIOR DAY	WEEK	WITHIN WEIGHT GROUPS: UNDER 500	500 TO 999	1000 TO 1999	2000 TO 4999	OVER 5000	TOTAL SHIPPED: LBS	M UNITS	M $	PER SHIPMENT AVG: LINES	LBS	$$$
	ZIP TOTAL			815	3	10	2		3	5	5				4031	161.9	10.9	10.6	403	1111
3478000007	BENEDECKS	45	CL	816	1	2			1	2					242	10.1	.6	4.0	121	321
3478000007	BENEDECKS	45	CL	816		3	1		1	3					745	24.1	1.6	5.6	248	548
3478000007	BENEDECKS	45	CL	816	1	3				3					600	25.5	1.8	9.3	200	629
	ZIP TOTAL			816	2	8	1		2	8					1587	59.7	4.0	6.6	198	522
DENVER, COLORADO	DEMAND AREA TOTAL				56	201	29	12	61	97	45	41	18		155856	6454.5	458.0	13.4	775	2290
6750000009	MORELAND WHOLESALE	45	WY	820	1	3	1		1	1	2				1806	86.5	5.6	17.3	602	1879
6750000009	MORELAND WHOLESALE	45	WY	820		3	2		2	2	1				1149	44.0	3.2	12.0	383	1087
6750000009	MORELAND WHOLESALE	45	WY	820	1	7	4		4	6	1				1930	82.5	5.6	9.2	275	810
8540000000	VANS CANDY & TOBAC	45	WY	820	1	5				2	1	2			4002	169.4	11.9	17.4	800	2389
8540000000	VANS CANDY & TOBAC	45	WY	820		4	1		1	1	2	1			2892	84.3	7.1	11.5	723	1788
8540000000	VANS CANDY & TOBAC	45	WY	820	1	5				2	2	1			3150	131.2	9.3	15.0	630	1874
	ZIP TOTAL			820	4	27	8		8	14	9	4			14929	597.9	42.7	13.3	552	1594
9058000002	WYOMING WHLSE DIST	45	WY	823		1				1					96	3.8	.3	7.0	96	310
9058000002	WYOMING WHLSE DIST	45	WY	823		2	1		1	2					123	5.2	.3	5.5	61	174
9058000002	WYOMING WHLSE DIST	45	WY	823		3				3					216	9.2	.6	5.0	72	220
	ZIP TOTAL			823		6	1		1	6					435	18.2	1.2	5.5	72	220
3910000003	CASPER DISTRIBUTIN	45	WY	826		1				1					293	12.5	1.0	15.0	293	1048
3910000003	CASPER DISTRIBUTIN	45	WY	826		1				1					223	11.7	.6	14.0	223	647
3910000003	CASPER DISTRIBUTIN	45	WY	826		1				1					470	21.7	1.5	23.0	470	1588
	ZIP TOTAL			826		3				3					986	45.9	3.1	17.3	328	1094
6300000004	M & F WHOLESALE CO	45	WY	827	1	1				1					60	4.0	.1	1.0	60	165
6300000004	M & F WHOLESALE CO	45	WY	827	1	1				1					152	8.0	.3	3.0	152	383
6300000004	M & F WHOLESALE CO	45	WY	827	1	1				1					75	5.0	.2	1.0	75	207
	ZIP TOTAL			827	3	3				3					287	17.0	.6	1.6	95	251
CHEYENNE, WYOMING	DEMAND AREA TOTAL				7	39	9		9	26	9	4			16637	679.0	47.6	11.5	426	1241
7404000007	W E PINCH	45	ID	835		1				1					362	12.2	1.3	5.0	362	1371
7404000007	W E PINCH	45	ID	835		2				2					824	30.5	3.1	7.5	412	1565
7404000007	W E PINCH	45	ID	835		1				1					320	11.2	1.1	6.0	320	1166
	ZIP TOTAL			835		4				4					1506	53.9	5.5	6.5	376	1417
5006000003	GLASER BROS 73-222	45	ID	837	4	8	4		4	8					870	43.8	2.7	3.2	108	340
5006000003	GLASER BROS 73-222	45	ID	837	8	12	10		11	10	2				3010	96.3	7.6	4.0	250	637
5006000003	GLASER BROS 73-222	45	ID	837	3	7	3		5	6	1				1735	78.3	5.1	4.2	247	738
5449000644	HEMENWAY & MOSER C	45	ID	837	2	6	2	1	2	5	1				1431	63.8	5.0	7.8	238	833
5449000644	HEMENWAY & MOSER C	45	ID	837		5	3		3	5					1466	67.2	4.9	6.8	293	997
5449000644	HEMENWAY & MOSER C	45	ID	837	3	4	3		3	4					917	43.0	3.0	4.2	229	774
	ZIP TOTAL			837	20	42	25	1	28	38	4				9429	392.4	28.3	4.8	224	681

Exhibit 5-2

Exhibit 5-3 provides a profile of the total shipment activity as detailed in Exhibit 5-2. It may be seen, for example, that out of a total of 12,861 shipments, there were 3,592 instances where two separate orders were shipped to the same customer on two consecutive days. The opportunity for savings through consolidation of freight shipments becomes obvious.

E. REDUCING ORDER PROCESSING TIME

Reducing customer order cycle time * may be used to improve customer service (with the ultimate objective of increasing sales), or it may be used to capture internal savings as was seen in the previous section of this chapter. This section shows how to analyze and reduce order cycle time in the *order processing* area.

Exhibit 5-4 shows a flow chart of a partially computerized order processing system. Note that the *average* time to completely process an order is nine days. Exhibit 5-5 shows the total distribution of time to ship orders (see dotted line). Note that while the average time is nine days, some orders required 20, 25 and more days to ship.

Exhibit 5-6 shows a proposed order processing flow which reduces the time to process an order from nine to five days. The reduction in time is accomplished in the following ways:

(A) Mail was received in a corporate mailroom which then redistributed mail to the various companies. *Time saved = 1 day.*

(B) Computerized manual features of present order processing method. *Time saved = 2.5 days.*

(C) Change from one-shift to two-shift pick-and-ship operation. *Time saved = 0.5 days.*

The proposed order processing system was introduced during a period when the company was experiencing heavy competitive pressures. Following the introduction of the new system, sales increased 2%. On an annual sales volume of $200 million, this amounted to $4 million per year!

F. SOME OPERATIONAL CONTROL REPORTS DESIGNED TO HELP YOU KEEP THE LID ON MAJOR WAREHOUSE COSTS

This section presents three reporting formats which have proven to be big dollar savers to warehouse managers. Of course, the reports merely organize important data. Savings are obtained only when the manager follows through and applies creative thinking to the respective areas.

First, Exhibit 5-7 presents an Overtime Control Report. The left-hand column shows the departments along with respective cost center numbers. To the right are the latest three monhs of forecasted and actual overtime hours by cost center. The current month shows only the forecast for overtime hours.

* Time from which a customer mails or otherwise communicates his order until that order is received by the customer.

```
V-000                        ORIGIN     REPORT COVERING SHIPMENTS FROM 08/20/   THRU 01/20/                          1000107-02

                                         * * * N U M B E R O F S H I P M E N T S * * *
   C U S T O M E R          ORIGIN DEST                             WITHIN WEIGHT GROUPS
                                         3 OR         EXTRA         500 1000 2000
 NUMBER            NAME                  LESS     SAME PRIOR   UNDER  TO   TO   TO  OVER    **** TOTAL SHIPPED***    PER SHIPMENT AVG
                                  ST ZIP LINE TOT DAY DAY WEEK 500  999 1999 4999 5000       LBS      M UNITS  M $   LINES LBS    $$$
```

	ORIGIN TOTAL
3 OR LESS LINES	4,517.0
TOT	12,861.0
SAME DAY	3,592.0
PRIOR DAY	1,513.0
WEEK	5,476.0
UNDER 500	7,130.0
500 TO 999	2,607.0
1000 TO 1999	1,772.0
2000 TO 4999	968.0
OVER 5000	384.0
TOTAL LBS	12,321,683.0
M UNITS	480,207.3
M $ SHIPPED	24,627.1
AVG NO OF LINES	10.1
AVG LBS	958.0
AVG $$$	2,802.5

Exhibit 5-3

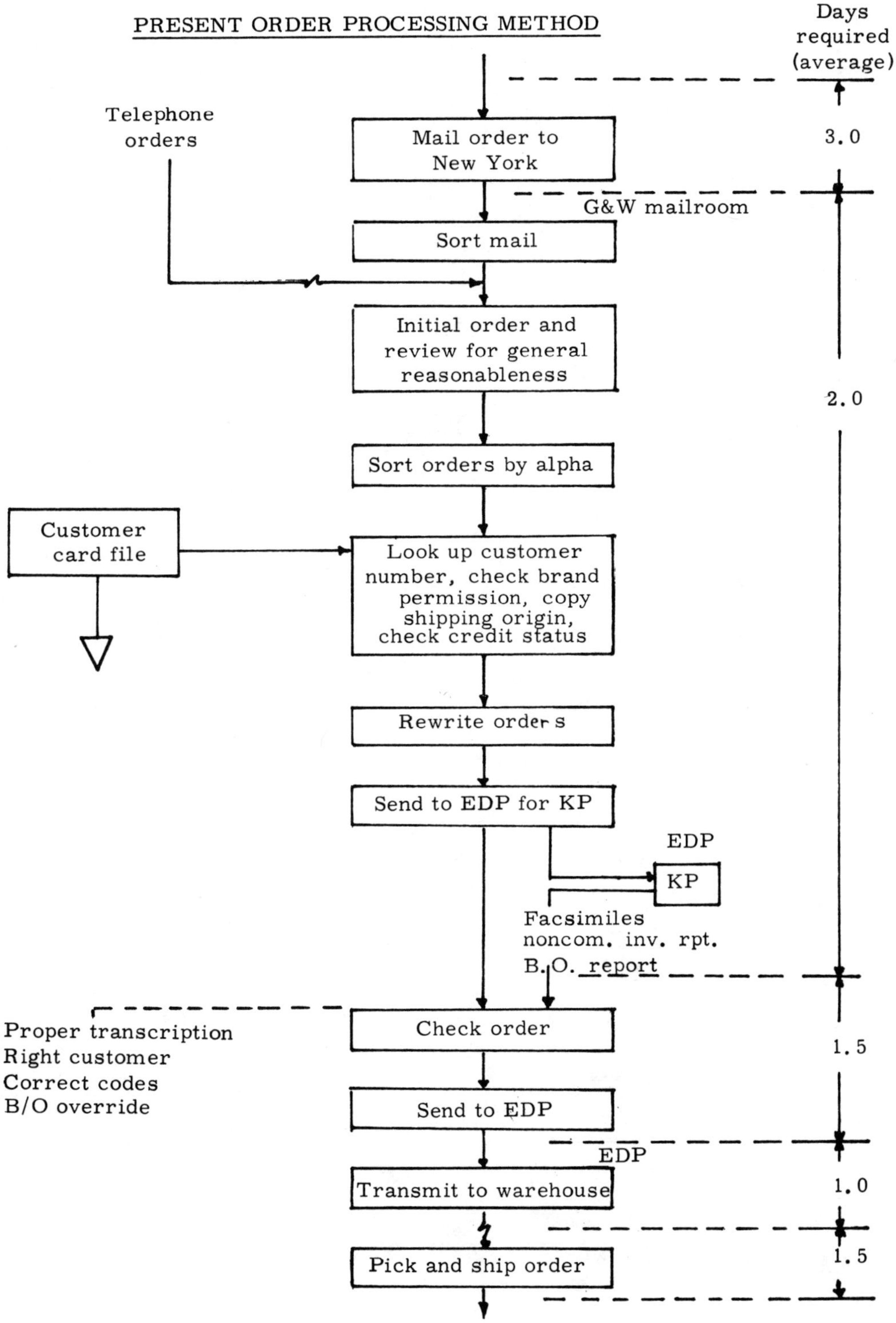
PRESENT ORDER PROCESSING METHOD
Days required (average)
Telephone orders
Mail order to New York
3.0
G&W mailroom
Sort mail
Initial order and review for general reasonableness
2.0
Sort orders by alpha
Customer card file
Look up customer number, check brand permission, copy shipping origin, check credit status
Rewrite orders
Send to EDP for KP
EDP
KP
Facsimiles
noncom. inv. rpt.
B.O. report
Proper transcription
Right customer
Correct codes
B/O override
Check order
1.5
Send to EDP
EDP
Transmit to warehouse
1.0
Pick and ship order
1.5

Exhibit 5-4

Distribution of Order Service Times[a]

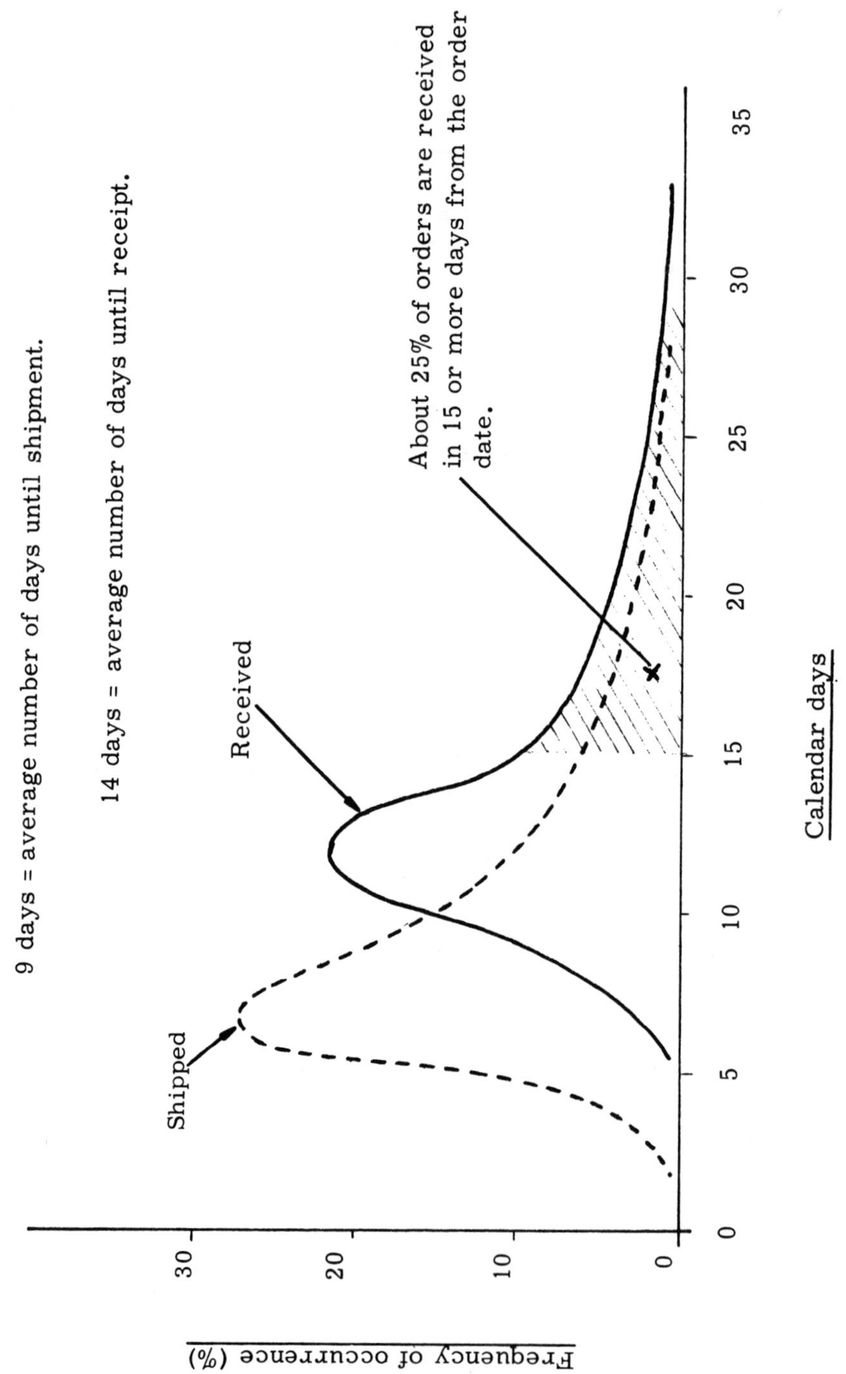

[a]All times are based on customer order dates.
Source: One-week sample of orders.

Exhibit 5-5

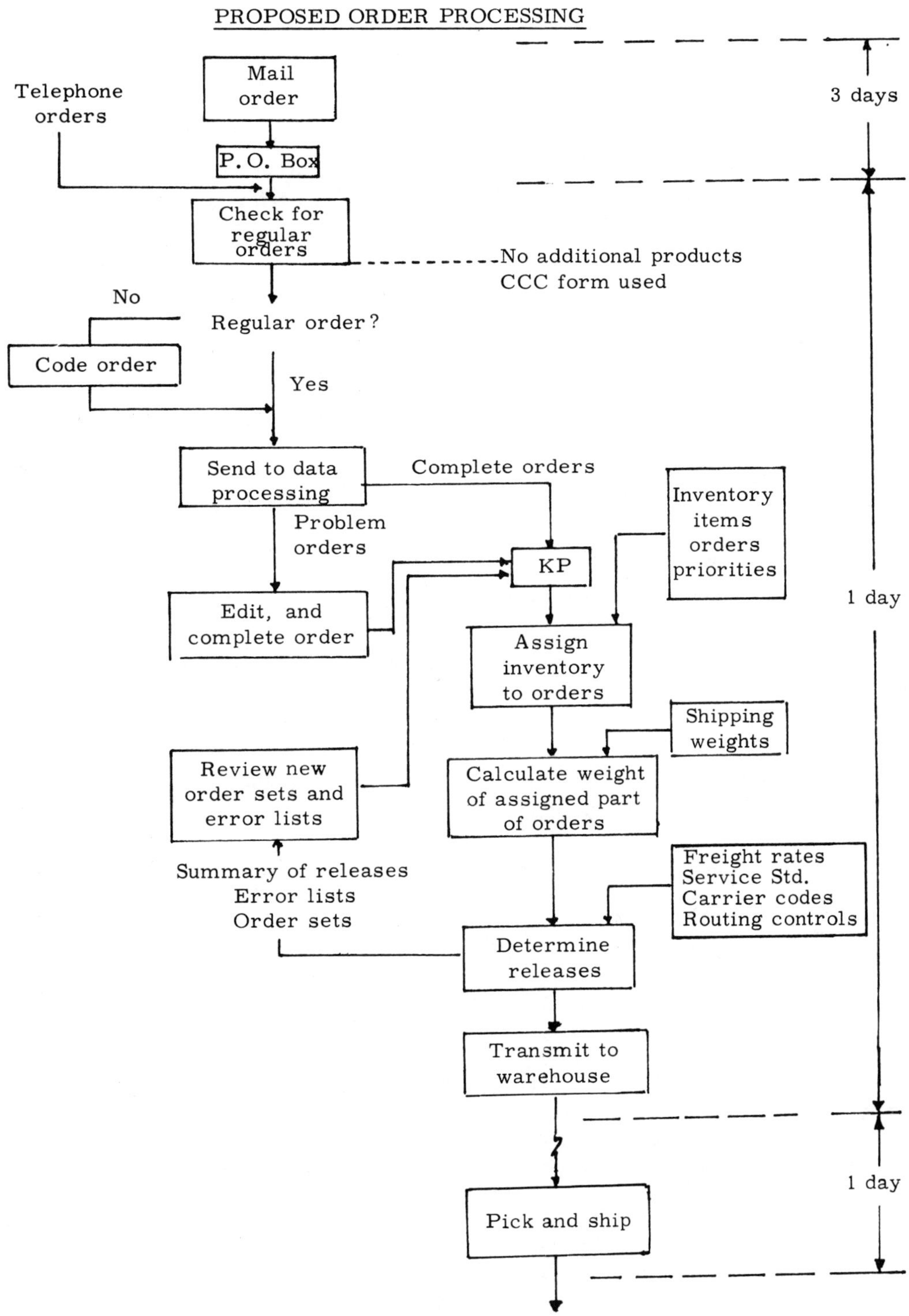
PROPOSED ORDER PROCESSING
Mail order
Telephone orders
P. O. Box
Check for regular orders
No additional products
CCC form used
Regular order?
No
Code order
Yes
Send to data processing
Complete orders
Problem orders
Inventory items orders priorities
KP
Edit, and complete order
Assign inventory to orders
Shipping weights
Review new order sets and error lists
Calculate weight of assigned part of orders
Summary of releases
Error lists
Order sets
Freight rates
Service Std.
Carrier codes
Routing controls
Determine releases
Transmit to warehouse
Pick and ship
3 days
1 day
1 day

Exhibit 5-6

SAMPLE
OVERTIME CONTROL
REPORT

Dept.	September F	September A	October F	October A	November F	November A	December F	December A
A. W. Deyo								
163	--	56	--	--	--	56	--	
624	288	388	360	558	300	672	360	
627	--	--	--	--	--	--	--	
628	8	9	8	11	8	6	8	
629	300	144	300	40	270	46	300	
632	300	60	300	8	300	8	300	
639	260	260	240	339	200	747	240	
697	200	52	200	--	200	--	200	
698	200	399	200	358	200	467	300	
Total	1556	1368	1608	1314	1478	2002	1708	
E. A. Reisel								
625	336	76	280	--	252	104	336	
630	--	24	16	--	--	--	--	
682	147	321	154	100	154	78	140	
683	360	412	360	24	360	96	450	
684	64	112	64	316	64	694	64	
685	300	552	300	452	300	236	300	
Total	1207	1497	1174	892	1130	1208	1290	
Grand Total	2763	2865	2782	2206	2608	3210	2998	

Exhibit 5-7

CASE IN POINT: A textbook distribution center manager reduced his overtime costs 20% when he took the time to estimate (forecast) future overtime requirements. This exercise led him to do a better planning job. He put certain cost centers on a two-shift basis; bought more productive materials handling equipment and used a computer to handle voluminous record keeping routines. Second, Exhibit 5-8 shows a Manpower Control Report on the same forecast vs. actual basis as the previous Exhibit. This report lists the total number of persons assigned to respective cost centers. Forecasts may change on a seasonal basis because of special promotions, or they may reflect planned growth or reduction in the work force.

CASE IN POINT: A cosmetics distributor found that his total labor force could be reduced 10% by creating a "labor pool." Each department had, over a period of time, hired sufficient personnel to cover *peak* work requirements. In creating the "labor pool," any persons not needed by a cost center on a given day were assigned to the warehouse manager, who would, in turn, reassign individuals to departments having a real need. Third, Exhibit 5-9 presents an Operating Expense Report. This report shows all labor, materials and overhead expenses charged to respective cost centers. The format needs no clarification. The report's main virtue is the feedback it provides on the individual manager's ability to control cost.

SAMPLE
MANPOWER CONTROL
REPORT

A. W. Deyo Dept.	September F	September A	October F	October A	November F	November A	December F	December A
*163	17	17	17	17	17	17	17	
*624	18	17	18	27(6,8,11)	25	25	25	
*627	14	14	14	14	14	14	14	
628	5	5	5	5	5	5	5	
*629	32	32	32	32	34	32	34	
*632	21	21(1)	21	13(7)	13	13	13	
**639	17	17	17	26(7,8)	31	30	31	
697	16	16	16	16(11,12)	16	18	16	
*698	24	23(2)	24	24(9)	24	24	24	
Total	164	162	164	174	179	178	179	
E. A. Reisel								
*625	19	19(4)	19	19	19	19	19	
630	2	2	2	2	2	2	2	
*682	25	25(1)	23	24(10)	23	24	23	
*683	21	21(3)	21	21	21	21	21	
*684	27	27(3&4)	26	25(8,10,11)	26	25	26	
*685	25	25(5)	25	25	23	25	23	
Total	119	119	116	116	114	116	114	
Grand Total	283	281	280	290	293	294	293	

*Figures include managers.
**Figure includes manager as of October.

1. Transfer of C. M. Killmer from D/682 to D/632-1 to replace T. Colaluca who went to D/634.
2. J. J. Galati, D/689 left company on September 19th.
3. Transfer of four in 683-2 to 684-2.
4. Transfer of J. P. Morgan, D/625 to D/684.
5. Transfer of A. M. Hebert, D/685 to D/163 to replace W. F. Sullivan who went to D/639-1. R. F. Haight, D/685 left for military service.
6. Transfer of J. P. Delahoyde, manager, and 10 people from Materials Planning to Materials Distribution, Dept. 624.
7. W. J. Freer, Dept. 639-1 left company October 31st. F. D. Malcher transferred to Dept. 865 on October 19th. Six people in Dept. 632 transferred to Dept. 639-1.
8. Transfer of J. B. DiCambio, former manager of Dept. 624 to 639-1. Transfer of C. A. Brower from Dept. 624 to 639-1 on October 1st. A. McSween transferred from Dept. 684 to 639-1.
9. Transfer of N. Morrone from Dept. 685 to 698.
10. Transfer of P. Yeaple from Dept. 682 to 687 on October 16th.
11. Transfer of C. Dodge from Dept. 684 to 697 on October 3rd, and T. Ponesse from Dept. 684 to 624 on October 3rd.
12. A. G. Davis transferred to Dept. 643 on October 3rd.

Exhibit 5-8

SAMPLE
OPERATING EXPENSE
REPORT

Department	August Difference Yr. to Date	September Budget	September Actual	Difference	September Difference Yr. to Date
A. W. Deyo					
163	-$ 2,497	$ 9,174	$ 8,527	-$ 647	-$ 1,044
624	-$ 1,838	$ 9,791	$ 9,991	+$ 200	-$ 1,638
627	-$ 9,615	$ 75,810	$ 63,871	-$ 11,939	-$ 21,554
628	-$ 5,299	$ 5,911	$ 5,121	-$ 790	-$ 6,089
629	-$ 21,700	$ 36,800	$ 24,469	-$ 12,331	-$ 20,031
632	-$ 3,500	$ 15,127	$ 13,262	-$ 1,865	-$ 5,365
639	-$ 9,329	$ 11,811	$ 10,738	-$ 1,073	-$ 10,402
697	-$ 26,757	$ 21,912	$ 14,515	-$ 7,397	-$ 34,154
698	+$ 80,902	$174,131	$181,807	+$ 7,676	+$ 88,978
Total	+$ 367	$360,467	$332,301	-$ 28,166	-$ 11,299
E. A. Reisel					
625	-$ 2,844	$ 12,359	$ 11,300	-$ 1,059	-$ 3,903
630	-$ 1,195	$ 1,763	$ 1,181	-$ 582	-$ 1,780
682	-$ 1,020	$ 46,275	$ 40,532	-$ 5,743	-$ 6,763
683	+$ 2,264	$ 19,421	$ 14,631	-$ 4,790	-$ 2,526
684	-$ 580	$ 15,538	$ 16,010	+$ 472	-$ 108
685	+$ 33,280	$ 30,741	$ 34,458	+$ 4,717	+$ 37,997
Total	+$ 29,905	$126,097	$118,112	-$ 6,985	+$ 22,917
Grand Total	+$ 30,272	$486,564	$450,413	-$ 30,151	+$ 11,618

September Difference - .07 1/4%

Exhibit 5-9

6

Selecting Materials Handling Equipment to Suit Your Particular Needs

USING "MATRIX COST ANALYSIS" TO SELECT THE MOST SUITABLE EQUIPMENT AT THE LEAST COST

The large variety of materials handling equipment available today often creates a bewildering set of alternatives for the Distribution Center manager. The matrix analysis presented in Exhibit 6-1 is designed to assist the user in narrowing his selection to the most logical choices.

In a later section of this chapter entitled "How to Select the Proper Materials Handling Equipment," a detailed approach to select one type of equipment over other alternatives is shown.

Often, equipment selection will be greatly influenced by specific use requirements, facility limitations, maintenance expertise required and available to keep the equipment operable and many other factors. Checklists included in this chapter are designed to aid you in not overlooking vital criteria in equipment selection.

The disastrous effect of not checking out equipment specifications against requirements may be seen in the following example. A consumer products manufacturer decided to open a satellite warehouse adjacent to a major port facility. A sophisticated automated order selection system was installed at a cost of $500,000. Annual savings of this system over a more conventional rack and fork-lift arrangement was estimated at $300,000 per year. After two years, the operation had to be shut down. The openings in the automated racks and selectors were fixed and were based on carton dimensions being used at the time the equipment was being designed. Over the 24-month period, carton dimensions changed. The fixed aperture system could not accommodate the new sizes and redesign costs were prohibitive. Lack of attention to all possible design criteria resulted in the scrapping of an otherwise sound idea.

The matrix analyses shown below will evaluate the following handling methods:

SUMMARY MATRIX

MATERIALS HANDLING EQUIPMENT SELECTION CHART

	Heavy Volume Over 600 Units/Load	Medium Volume 300-600 Units/Load	Light Volume Under 300 Units/Load
Short haul 0-200' each way	Tractor	Tractor-Train	Tractor Train Fork Lift
Medium Haul 200'-800' each way	Monorail Tractor-Train Conveyor Towline	Tractor Train	Tractor-Train
Long haul Over 800' each way	Conveyor Monorail Towline	Conveyor Monorail Towline	Tractor-Train

Approximate first costs varations for a minimum installation under these same classifications are as follows:

	Heavy Volume	Medium Volume	Light Volume
Short haul 0-200'	$ 6,000-10,000	$ 6,000-10,000	$ 6,000-10,000
Med. Haul 200'-800'	6,000-145,000	6,000-10,000	6,000-10,000
Long Haul over 800'	168,000-570,000	168,000-570,000	6,000-10,000

Exhibit 6-1

- Electric fork lift
- Gas fork lift (gasoline, diesel and LPG)
- Towline (infloor and overhead)
- Conveyor
- Monorail
- Electric tractor-train
- Gas tractor-train (gasoline, diesel or LPG)
- Electric tractor-train—driverless

When evaluating which long haul transport method to employ, the prime considerations are the number of loads to be moved per day and the transport distance. The criteria shown in Exhibit 6-1 will determine the area of consideration. Exhibit 6-2 shows other continuing cost relationships which must also be considered.

CONTINUING COSTS TO CONSIDER

1. Space Requirements
2. Operator Costs
3. Staff Other Than Operator
4. Maintenance Time & Cost

	Tractor Train	Monorail	Conveyor	Towline	Fork Lift
Space	High	Low	Low	Med.High	Low
Operator Min/Unit load	Short Haul .15 / Med. Haul .55 / Long Haul 1.12	none	none	none	1.1
Staff Min/unit load	2.6 min.	1.5 min.	1.5 min.	2.6 min.	.92 min.
Maint. Time & Cost	High	Low	Medium	Low	High

Exhibit 6-2

Other criteria which might present themselves in specific applications are:

1. Clear aisles
2. Cross-traffic through transport aisles
3. Staffing requirements
4. Maintenance time and costs
5. Floor space requirements
6. Building under construction
7. Indoor and outdoor applications

CASE IN POINT: The Summary Matrix Equipment Selection Chart shown in Exhibit 6-1, was used in evaluating an independent school district warehousing complex having onfloor storage and ramps for hauling between buildings. The study showed that racking and the use of fork-lift trucks was most suitable and would save $35,000 per year in labor costs. Volume was light and hauling distance was under 200 feet each way. Theoretically, straddle fork trucks, which require minimum aisle widths, would have been the most efficient units to buy. Minimum aisle widths mean more racking in a given area and greater storage efficiency. However, straddle trucks could not negotiate the ramps necessary for intrabuilding transfer. Counterbalanced fork trucks were, therefore, specified.

CASE IN POINT: In the case of a soft-goods distributor, the Summary Matrix Analysis showed that either tractor trains or fork-lift trucks would be suitable (light volume and short hauls). However, as indicated in Exhibit 6-2, consideration must also be given to space requirements and other factors in deciding which alternative to use. Two space factors resulted in fork trucks finally being selected. First, fork trucks could operate in narrower aisles, requiring less building space. Second, both seasonal and workshift traffic surges required temporary use of aisles for staging pallet loads going into stock. This would be possible if fork trucks were used, but fixed-path tractor-trains do not provide the necessary flexibility. Additional staging areas which would be used for only brief periods would have been necessary with tractor-trains.

The selection of fork trucks resulted in a space saving of 15,000 square feet. 15,000 square feet @ $10 = $150,000 savings in building costs.

Staffing requirements covering the various modes of materials handling transport are presented in Exhibit 6-3. Labor cost comparisons weigh heavily in the economic analysis and justification for more automated approaches to materials handling.

Given the general and individual parameters shown below, Exhibits 6-4 and 6-5 (pages 118 and 119) present cost and capacity planning data respectively.

GENERAL PARAMETERS

1. All flow is from point "A" to point "B".
2. Return of empties included where applicable.
3. Units measure 3'x3'x3' with weight of approximately 2000#.

4. Transport will be required equally throughout the workday.
5. All cycle times are based on 400 working minutes per eight-hour day.
6. Loading and unloading costs have not been calculated for this study.

INDIVIDUAL PARAMETERS

A. *TOWLINE STUDY*
Units are placed on towline carts prior to delivery to dispatch station, and carts are removed from delivery station still loaded.

B. *MONORAIL STUDY*
Unit loads are delivered to monorail dispatch station at correct height to be placed on monorail without mechanical assists. The unit loads will be deposited at monorail discharge station at discharge height.

C. *CONVEYOR STUDY*
Unit loads are delivered to conveyor dispatch point at the correct height to be placed on conveyor without mechanical assists. The unit loads will be deposited at conveyor discharge point at height of conveyor.

STAFFING REQUIREMENTS
(Minimum Number of People Needed to Operate)

A	B	C
Electric Fork Lift (O) Gas Fork Lift (O)	Towline (S) Conveyor(S) Monorail(S) Tractor-Train Driverless (S)	Tractor - Train (O)(S) (Gas & Electric)

(O)=Operator - Required 8 hours per day.

(S)=Support - This job is to load and unload transporter at each terminus. This requires two men due to the distance between job sites. These jobs may not require 8 hours per day.

Exhibit 6-3

INITIAL COST EVALUATION PER FOOT

If Distance from A to B is	Electric Fork Lift W/Battery	Towline & Carts	Gas Fork Lift	Monorail	Conveyor	Gas Tractor Train	Electric Tractor Train
Per Foot	N.A.	Towline Spurs 4 @ 500-280 Ft.	N.A.	$90.00 Turn - 10 Each	$125.00	N.A.	N.A.
100'	$10,000	$14,000 Carts 16 @ 200-3,200	$7,000	$19,800	$12,500	Tractor-$3,000 Carts $ 2,400 (Cts.-16 @ 150)	Tractor-$4,000 Carts $2,400 (16 @ 150)6,400
200'	10,000	19,000 Carts 16 @ 200-3,200	7,000	37,800	25,000	5,400	6,400
400'	10,000	29,000 Carts 16 @ 200-3,200	7,000	73,800	50,000	5,400	6,400
800'	10,000	49,000 Carts 16 @ 200-3,200	7,000	145,800	100,000	5,400	6,400
1,600'	10,000	89,000 Carts 16 @ 200-3,200	7,000	289.800	200,000	5,400	6,400
3,200'	10,000	168,600 Carts 16 @ 200-3,200	7,000	577,800	400,000	5,400	6,400

Exhibit 6-4

CAPACITY IN UNIT LOADS PER 8-HOUR (1) DAY
(For a Minimum Installation)

If Distance From "A" to "B" is (2)	Electric Fork Lift (1 load)	Electric Fork Lift Stacked(3)	Gas Fork Lift Stacked(3)	Towline	Monorail	Conveyor	Manual or Automatic Gas Tractor Train(4)	Manual or Automatic Electric Tractor Train(4)	Electric Fork Lift w/trailers(5)
100'	274	300	330	800	800	800	(262 trips) 2096	(206 trips) 1648	(185 trips) 1665
200	198	230	292	800	800	800	(229 trips) 1832	(160 trips) 1280	(137 trips) 1233
400	127	160	242	800	800	800	(174 trips) 1492	(109 trips) 872	(91 trips) 819
800	74	100	182	800	800	800	(117 trips) 936	(675 trips) 540	(54 trips) 485
1600	41	56	120	800	800	800	(705 trips) 564	(39 trips) 292	(30 trips) 270
3200	20	31	72	800	800	800	(45 trips) 360	(11 trips) 88	(16 trips) 144

(1) - 400 working minutes.

(2) - Return of empties calculated where required.

(3) - Requires 4000# lift.

(4) - Tiering on carts not considered - 8 load trip.

(5) - Tiering on carts not considered - 9 load trip (one on forks).

Exhibit 6-5

D. *TRACTOR-TRAIN STUDY*
Unit loads are placed on train carts prior to delivery to dispatch area and carts are removed from discharge area still loaded.

E. *FORK-LIFT STUDY*
Units are palletized prior to delivery at dispatch area (one high) and deposited at discharge area on pallets (one high).

Having the matrix analysis shown in the above Exhibits, the warehouse manager will be able to narrow his choice to a few logical alternative materials handling methods.

HOW TO SELECT THE PROPER MATERIALS HANDLING EQUIPMENT

Assume that through use of the matrix analysis, you have narrowed your equipment alternatives to the following:

Fork Trucks

Tractor-Trailer Trains—Driver-Operated

Tractor-Trailer Trains—Driverless

Towline Equipment

Exhibits 6-6, 6-7, 6-8 and 6-9 show the respective economic analysis needed for individual equipment selection.

Exhibit 6-6

COMPARISON OF INTERPLANT MATERIALS HANDLING EQUIPMENT

FORK TRUCK

Assumptions

Average moves 500 feet in or out or 1,000 feet. At 3 mph (264 fpm), travel time = 4 minutes, including picking up and dropping pallets.

At peak, must handle 219 pallets in and 123 pallets out per hour. Sufficient equipment must therefore be provided to handle 219 pallets in. Peak will last six months.

Fork trucks will carry two pallet loads at a time; therefore, will make 110 round trips; say 150 to allow for times when two pallet loads cannot be handled.

150 trips at 4 vehicle minutes per trip = 600 vehicle minutes/hour. At 60 minutes/hour = 10 vehicles at peak. Purchase ten (10) fork trucks for full-year operation. As an alternative, five (5) could be purchased and the remaining five (5) rented during peak 6-month period, if accounting deemed this advisable. Purchase price of LP truck @ $9,500. There would be five (5)

full-time employees, with five (5) part-time employees hired for the peak 6-month period. Operator cost @ 500 per month.

Costs

	Fixed(1)	Annual
10 fork trucks @ $9,500—Purchased	$95,000	
Full-time operators—60 man-month @ $500/mo.		$ 30,000
Part-time operators—30 @ $500/mo.		15,000
Total first year costs =	$95,000	$ 45,000
Total cost after 3 years =	95,000	135,000
Total cost after 5 years =	95,000	225,000
Total cost after 10 years =	95,000	450,000

Notes—(1) Assuming maintenance costs equal in all alternates.

Exhibit 6-7

TRACTOR-TRAILER TRAINS—DRIVER-OPERATED

Assumptions

Average moves 500 feet in and out or 1,000 feet. At 3 mph 264 (fpm), travel time = 4 minutes to hitch and unhitch trailers, or 15 minutes per roundtrip, or 4 trips per hour.

At peak, must handle 219 pallets in and 123 pallets out per hour. Sufficient trains must be provided to handle 219 pallets in. Peak will last six months.

Train will carry six (6) pallet loads; therefore, 219 ÷ 6 = 37 trailer trips must be made each hour. At 4 trips per hour, require 10 tractor-trailer trains at peak; say 12 to allow for inefficiencies.

Purchase twelve (12) tractors @ $3,000 and all trailers (100 @ $175 = $17,500). Six (6) tractors would be used continuously throughout year and six (6) tractors would be used only during 6-month peak period. Six (6) full-time operators and six (6) part-time operators would be required at $500 per man month.

Costs

	Fixed(1)	Annual
12 tractors @ $3,000—Purchased	$36,000	
100 trailers @ $175—Purchased	17,500	
Full-time operators—72 man months @ $500		$ 36,000
Part-time operators—36 man months @ $500		18,000
Total first year costs =	$53,500	$ 54,000
Total cost after 3 years =	$53,500	$162,000
Total cost after 5 years =	53,500	270,000
Total cost after 10 years =	53,500	540,000

Notes—(1) Assuming maintenance costs equal in all alternates.

Exhibit 6-8

TRACTOR-TRAILER TRAINS—DRIVERLESS AND FULL TRAILERS

Assumptions

Average moves 500 feet in and out or 1,000 feet. At 1.5 mph (100 fpm), travel time = 10 minutes, but allow 5 minutes to automatically unhitch and manually hitch up trailer trains, or 15 minutes per round trip, or 4 trips per hour.

At peak, must handle 219 pallets in and 123 pallets out per hour. Sufficient trains must be provided to handle 219 pallets in. Peak will last six months.

Train will carry six (6) pallet loads; therefore, 219 ÷ 6 = 37 trailer trips must be made each hour. At 4 trips per hour, require 10 tractor-trailer trains at peak; say 12 to allow for inefficiencies.

Purchase twelve (12) tractors @ $10,000 and all trailers (100 @ $175 = $17,500). Cabling, with controls to permit stations to "call" for tractor and for traffic control on the various loops, estimated at $20 per foot. Assume path similar to towline, or 3,200 feet.

Costs

	Fixed(1)
12 Driverless Tractors @ $10,000	$120,000
100 Trailers @ $175	17,500
Cable—3,200 feet @ $20 per foot	64,000
Total Cost =	$201,500

Notes—(1) Assuming maintenance costs equal in all alternates.

Exhibit 6-9

TOWLINE EQUIPMENT

Assumptions

Three interconnected towline loops are proposed. Operating at 60 to 80 fpm, these systems provide sufficient capacity for all projected requirements. When properly maintained, the towline will operate unattended and move coded carts to designated spurs throughout the facility.

If the towline were put in at a later date, add 25% to equipment cost. This does not include extra costs that would be incurred initially to design the building for this eventuality.

If the towline track were installed only, deduct about 15%.

Costs

The towline equipment required for the current volumes, which in addition are capable of handling future demand requirements, are as follows:

—Perimeter towline	$149,600
—Storage area	77,800
—Assembly	95,300
—Contingency	32,300
Sub-total	$355,000
—Initial cost of additional work to concrete floor	48,000
Total	$403,000

CONCLUSIONS

As can be seen from the comparisons, all equipment has a lower first cost than the towline. However, in the long run, the total cost of the fork truck and driven or driverless tractor-trailer train alternative will doubtlessly be substantially greater than the towline.

Given a limited budgetary situation, the most prudent decision might be the driverless tractor-trailer train operation. Even though the long-range cost will be greater, this equipment will require a smaller initial investment, provide a great deal of flexibility (paths can easily be changed), can be pulled or moved out without leaving a major component behind and can easily be added to as volumes increase.

It should be noted, however, that for new construction, once this alternative is chosen, it will be highly unlikely that an in-the-floor towline can ever be installed in the building.

MATERIALS HANDLING EQUIPMENT—A COST SAVING REVIEW OF THEIR ASSETS AND LIABILITIES

Assets and liabilities of four of the more popular forms of materials handling equipment are listed below:

FORK-LIFT ANALYSIS

(Electric-in-Plant)

(Gas-Outdoor)

ASSETS	LIABILITIES
1. When stacking problem arises at either terminus, this equipment can perform all functions	1. Requires trained operator
	2. With electric equipment, when floor height variations become significant, the battery size might become prohibitive
2. Complete flexibility	
3. No staging area required	3. Tiering is required to handle more than one load at a time
4. Does not require very wide aisles	

SUMMARY

This method of horizontal transport is most desirable when unit loads must be tiered or untiered at each terminus. This one piece of equipment (and one operator) can perform the work of three pieces of equipment.

1—unstack
2—move
3—restack

No permanent installations such as tracks, guides or posts are required.

TRACTOR-TRAILERS

(Electric-in-Plant)
(Gas-Outdoor)

ASSETS

1. Large number of units per cycle
2. No permanent installation required
3. Flexibility within wide aisles

LIABILITIES

1. Requires trained operator
2. Service between stations is intermittent
3. Requires one man at each terminus to handle load
4. On electric equipment substantial variances of floor elevations (ramps) will make battery size prohibitive
5. A staging area is required (sample problem requirement is 1,220 sq. ft.)

SUMMARY

This is an excellent mode of transporting over the "long haul" when no installations are desired. It is a manual system in that it requires a driver to operate the tractor. This system has the versatility to operate in any unblocked wide aisle, but must be provided with a staging area at each terminus.

DRIVERLESS (AUTOMATIC) TRACTOR-TRAINS

ASSETS

1. Requires no driver

LIABILITIES

1. Must operate on fixed route
2. Requires permanent installation
3. Requires one man at each terminus to handle loads
4. On electric equipment, substantial variables of floor heights (ramps) will make battery size prohibitive
5. Service between stations is intermittent
6. Staging area is required (sample problem requirement is 1,220 sq. ft.)

SUMMARY

When a permanent route is required, driverless equipment is very practical. It requires either a channel in the floor or a buried wire.

CONVEYOR

ASSETS

1. No return empty (dead-head) cycle required
2. Operator required
3. Can be adapted to grades
4. Can accommodate high volume
5. Does not require wide aisles
6. Continuous operation

LIABILITIES

1. Conveyor line cannot be crossed with any equipment (other than monorail)
2. Completely rigid—offers no flexibility
3. Requires installation
4. Requires operator at each terminus to load (unload) (This may not require 16-man/hr./day)

SUMMARY

This is the most rigid of systems discussed. Aisles *cannot* be run across the conveyor line. This goes in one direction only, so that a second (return) aisle will not have to be created. This system should only be considered when a large-volume move is required. Gates or in-floor conveyors may be added at extra expense to cross aisles.

7

Effective Warehouse Layout Planning Reduces Materials Handling Costs

A. OBJECTIVES IN DESIGNING A NEW WAREHOUSE

1. The Place of the Warehouse in a Cost-Effective Distribution System

The average management opinion is that shipments should be made direct, with a minimizing of rehandling and warehousing. This attitude is not disappearing, as is evidenced by the increasing use of electronic data information flows between Sales, Production and Distribution to speed up deliveries.

Although speedup information may minimize warehousing requirements in many cases, there is no evidence of the elimination of warehousing.

The warehouse, to Materials Management men, performs a multiple service:

(1) To balance production; e.g., carbon black—steady production with irregular shipping demands.

(2) To balance sales demands; e.g., holiday items and sales promotions in the cosmetic fields—short cycle, with heavy shipping surge setting demand well over normal production capacity.

(3) To provide prompt or immediate service; e.g., bearings, pump parts and other components—emergency service to the customer.

(4) Insurance warehousing; e.g., paper machine gear train—items stocked by the customer to minimize down-time due to breakdown. Not normally stocked by manufacturer.

(5) Unplanned or dead warehousing; e.g., excess stocks due to poor analysis and planning by someone:

(a) Overruns potentially unsaleable—pop cans.
(b) R & M Parts—refinery trays.
(c) Old or off-spec material as a result of poor warehouse management—synthetic rubber.

The first four reflect planning and control; the fifth reflects a breakdown in one or all of the four.

The current trend today in industry is to minimize inventories of purchased items or materials by letting the producer have the responsibility of maintaining an adequate inventory level; e.g., a brewery and can manufacturer. This places great pressure on Materials Management in planned warehousing and deliveries, both at the customers' and shipper's facilities.

Current requirements call for expeditious service to meet customers' desires to operate with minimum inventories. Materials Management must then think in terms of optimum distribution. The ideal situation would have no warehousing. Distribution costs would be minimized.

Materials Management is faced with the job of approximating the economics of direct shipments and at the same time producing adequate service. Two factors must be considered:

(1) Warehousing locations: The ideal condition here would be a consigned stock in each customer's plant. The public utility of the City of St. Louis has consigned stock of lube oils in each generating plant. As a drum is issued, a copy of the requisition is sent to the supplier who invoices the customer. The supplier maintains the required inventory level.

Most suppliers cannot afford to operate under such conditions. The inventory investment alone would make the practice uneconomical.

The next step then is to set up the distribution points. There must be the following considerations:

(a) Service to the customer.
(b) Economy of distribution.

The second is a very decisive factor because service must be subject to economy. This is a factor which will vary with conditions. For example, a metals distributor warehoused three items in Akron for some years. Recently, a study revealed that one product could be distributed more economically from Cincinnati. Since this company uses commercial warehouses, economy of transportation was the deciding factor. Construction of a new facility was not necessary.

(2) The internal operation in the warehouse: Speed of delivery will be an important factor. Increasing pressure is brought toward the shortening of the cycle between the receipt of the order and the shipment of the goods. For example, an electronics equipment manufacturer has developed to a high degree, speed of information flow between the sales office and the distribution warehouse. This is a highly sophisticated system and, naturally, will not apply to all situations. This, then, puts the onus on you to analyze those conditions peculiar to your operation and increase the speed of information flow.

Along with speed of delivery, company management will also hold you responsible for an economical operation in terms of labor and space utilization. Anyone familiar with commercial warehousing will be aware that optimum space utilization is a major factor affecting the profit picture. The following layout and layout-related items should be considered:

(a) *Labor*—Use mechanical equipment to minimize manual handling and movement.

1) At the receiving areas: conveyors, palletizers, fork-lift trucks, drag line carts, other mobile equipment.

2) Transport into and out of storage: conveyors, fork-lift trucks, stack cranes, monorails, drag lines, tractor-trailer trains, transfer equipment, bridge cranes, other mobile equipment.

3) At the accumulation and shipping area: conveyors and chutes, both transport and storage, fork trucks, other mobile equipment, bridge cranes.

(b) *Space*—Use storage equipment designed for specific items and optimum space utilization.

1) Pallets for unit loads of one item which can sustain the weight of a stack of optimum height. This will also apply to unpalletized unit loads—canned goods, bagged material, etc.

2) Box type pallets for unit loads of one item which cannot sustain the weight of the stack at optimum height—soft goods packaged but not cased.

3) Pallet racks for single utilized pallets of a single item or palletized irregular shapes which are not stable supports.

4) Racks for horizontal storage of long shapes not adaptable to palletization rods and bars, rolled carpeting, box sheets.

5) Racks for vertical and/or handling storage of items of peculiar characteristics—pump impellers, large gaskets.

6) Bins, shelving, stack bins, etc. for the storage of small items. There are a variety of styles, most of which are well known.

There are other forms of storage in a distribution system—tanks, silos, hoppers, special cars, etc. All are subject to the same basic principle of optimum utilization of potential storage space.

2. Planning the Layout

This requires detailed study based on the previous discussion plus several other factors. The layout is predicted on the storage space requirements converted to storage area and servicing areas.

A. *Storage area*

This will be indicated in terms of:

1. Unitized lot sizes with adequate depth according to inventory volume and turnover.
2. Supplemental unit storage such as box pallets and pallet racks.

   NOTE: Either of the above could be active on backup stock for binned material.
3. Binned stock.
4. Racks, unitized long items and floor storage.

B. *Shipping and receiving areas*

The size and location of these are dependent on several factors:

1. Volumes handled.
2. Number of cars, trucks and trailers to be serviced daily.
3. The receiving and shipping accumulation requirements.
4. Packing area requirements.

C. *Service areas*

Offices, rest and locker rooms, security rooms, etc.

D. *Aisles*

They must be so laid out that prompt and free travel is possible from receiving to stores and back to shipping. Main travel aisles in a mobile truck operation must be wide enough for passing. Service aisles in the storage area may vary from approximately six to approximately 20 feet depending on the transport equipment.

3. Some Basic Principles to Remember

A. *Don't assume you have to build*

A paper products distributor, for example, was prepared to expand its 100,000 square foot warehouse by an additional 25,000 square feet. A thorough review showed this to be necessary, resulting in a savings of $250,000. Use of drive-in racks, better work methods, an improved stock locator system and other space utilization improvement ideas postponed the need for additional space for many years.

B. *Small size may be a virtue*

One firm decided to consolidate several geographically scattered warehouses into a single location, hoping to take advantage of economies of scale. Unfortunately, the product line did not lend itself to volume handling. Cost reductions did not materialize and transit time increased for many customers. As service levels fell and competitors moved in, the firm was forced to decentralize its operations again.

Many firms build too much space on speculation. Planning for expansion is fine, but be realistic. Develop a space requirements plan that will carry you out three to five years in the future. One company, experiencing new and rapid growth, built a 400,000 square foot warehouse that left 100,000 square feet idle for years. The business matured and the growth rate normalized.

C. *Study automation carefully*

A consumer goods manufacturer built a 90,000 square foot automated warehouse that became inoperable within 18 months. Changes in package sizes could not be accommodated by the order selection equipment. Two million dollars was lost.

Automation requires a great deal of pre-planning. You don't buy automated systems out of catalogs or have suppliers assemble them for you from off-the-shelf components. And the free engineering that may be provided is often worth as much as you pay for it. Usually, automated systems are necessarily custom designed, with equipment supplied by a variety of firms.

D. *Don't overlook your information systems*

Distribution centers won't operate efficiently without a well-planned information system. This may call for computer capability of either on-site time sharing or terminal access to a remote computer center. The computer can handle the following work:

1. Order processing.
2. Inventory management.
3. Preparation of shipping documents.
4. Order selection paperwork.
5. Management analysis reports.

B. THE STOCK LOCATOR SYSTEM

1. The Locator System

The locator system is the road map of the storage area. For large-lot unitized storage, whether serviced by a fork truck or a crane, the simple grid system is sufficient.

This allows optimum space utilization through use of random storage procedures.

Storage areas will have definite area limitations so that the stack depth is limited. Therefore, as an area is emptied, it may be filled again with large-volume items having the same area requirements, but not necessarily being the identical item.

In a normal operation with continual turnover, the warehouse supervisor or even the fork truck or craneman may make the selection. The location will be transmitted to either a locator file or incorporated with the data on the inventory card. When the goods are required, the location is noted on the issuing papers to the picker. This practice is basic in commercial warehousing.

Pallet rack storage may also be subject to the random storage pattern. An example is in the following pages. However, because of the small lots and less active movement it may also fall into the binning category discussed next.

Bins or broken case lots are usually subject to identifications by aisle and location in the aisle. Each aisle is numbered as well as each opening. Random storage is not possible because of the small-volume movement. Material in the bins will also be identified in the locator file. Several items of small size may be in one location.

As more sophisticated warehousing techniques are adopted, the control of the selection under the random system becomes more complicated. Feeding to, placing in and removal from storage become automatic; the personal element of the warehouse staff becomes superfluous. Selection of the storage space can be made at a control point, either automatically or semi-automatically. Removal will be made by the same method.

A major electronics components manufacturer adopted a random storage locator system and reduced warehousing requirements from 310,000 square feet to 110,000 square feet. The system, which cost approximately $1.5 million, enabled the company to consolidate warehousing of raw materials and finished goods under one roof. Previously, several satellite warehouses were used. The location of items in storage is controlled by a computer file and sorting system. The exact location of any item can be determined in seconds. A stacker crane operator then moves the crane to the proper location and removes the material mechanically. The payback period was less than one year.

2. Stock Rotation

This ties in with the locator system because lot location is the basis of adequate inventory rotation. The member of the organization who is responsible for determining the location for an item should be responsible for inventory rotation.

However, the ultimate responsibility for inventory rotation must rest on warehouse supervision. Periodic policing will be necessary.

The degree of complexity of rotation varies. The following are some examples:

1. *Large lots in unit loads.* If the item is so active that it is always stored in the same area, rotation is simple. The fork-truck operation can be directed to work from right to left in the area. There will always be an open area between the incoming and outgoing goods.

In cases where there is a fluctuation of item storage requirements, control of inventory rotation rests back at the location card file. As the requisition is released for picking, the locations of the oldest goods are entered.

2. *Flow racks or conveyors with gravity movement guaranty stock rotation of goods in that particular location.* The oldest goods are always at the picking end. Control must be exerted at the backup stock to move the oldest goods to the racks.

3. *Bins, shelves and racks containing small lots, or goods picked in small quantities, can best be controlled by manual relocation of the oldest goods to a place in front of newer stock. Again, backup stock is a controlling factor.*

Automatic warehousing, as in the locator system, provides a more precise control where it applies. As an item is delivered, it can be assigned a location upon receipt either from a locator file or from an electronic reservoir of open locations. When it is to be picked, the location is already on the issuing paperwork and the goods are released automatically or by a mechanical transporter.

An electronics manufacturer uses a system similar to the one above except that all goods are palletized at the packaging station and later sold in unit loads. Movement is by fork truck.

Each pallet load is numbered and stored by number. The product has a specification range acceptable to most of its customers. However, over a period of years, some customers have set up more rigid requirements within the range. A production run, then, which may cover up to two weeks, is controlled so that a certain part of the run will be in the narrower range. This is noted and entered in the electronic data system. When an order is received, the system will issue the shipping papers with pallet load numbers and the specification sheet. Normally, pallet selection is in numerical order. If the specification sheet calls for the narrower range, the corresponding pallet numbers are on the shipping paper. This offers selectivity as well as guaranteed rotation. If, toward the end of the month, the material in the narrow specification range has not moved, it is shipped as regular stock.

There are certain safeguards that are used to control rotation:

1. After policing, if old material is discovered it can be noted and installed for use.

2. The most common is numbering or coding. Motors and instruments may be serialized and the lowest number issued first. In the case of foodstuffs, chemicals, etc., a code number identifies the oldest material to guide the picker in moving the older goods.

C. STORAGE SYSTEMS

1. Calculating Storage Requirements

Exhibit 7-1 presents a format for calculating storage requirements for virtually all warehousing situations. Use of this format reduces the layout planning effort to a matter of days rather than, as often happens, weeks and even months.

The procedure is simply one of recording from inventory or purchasing records the peak, point-in-time quantities to be stored for each item in stock. If, for example,

STORAGE PROFILE ANALYSIS

	PRODUCT NO.	PRODUCT NAME	UNITS PER BOX	CARTON	AVERAGE SALES PER DAY	MONTH	ESTIMATED BOXES PER DAY	MONTH	YEAR	BOXES PER PALLET	PALLETS PER YEAR	SIZE OF BOX
1	4651	Celery Frost	110	-	7	134	.06	1.22	15	9	1/3	10x10x6
2	4652	Grey Mat	"	-	3	45	.02	.41	5	"	1/3	"
3	4654	Brown Mat	"	-	5	105	.05	.95	11	"	1/3	"
4	4657	Jade Frost	"	-	11	209	.10	1.90	23	"	1/3	"
5	4658	Aqua Frost	"	-	16	258	.11	2.35	28	"	1/3	"
6	4659	Brown Frost	"	-	4	72	.04	.66	8	"	1/3	"
7	4660	Grey Frost	"	-	4	71	.04	.65	8	"	1/3	"
8	4661	Blue Frost	"	-	7	141	.07	1.28	16	"	1/3	"
9	4663	Copper Frost	"	-	6	119	.06	1.08	13	"	1/3	"
10	4664	Snow Blue	"	-	7	126	.06	1.15	14	"	1/3	"
11	4665	Snow Violet	"	-	7	135	.06	1.23	15	"	1/3	"
12												
13	4701	Blue Smudge	288	-	4	72	.01	.25	3	36)		13x12x11
14	4702	Brown Smudge	"	-	3	60	.01	.21	3	)	1	"
15	4703	Burgundy Smudge	"	-	6	109	.02	.38	5	")		"
16	4704	Charcoal Smudge	"	-	3	56	.01	.60	3	")		"

Exhibit 7-1

17													
18	4801	Ebony	130	-	1	18	.01	.14	2	12)			18x18x15
19	4802	Charcoal	"	-	2	25	.01	.20	3	")			"
20	4803	Dark Tortoise	"	-	2	38	.01	.30	4	")	1		"
21	4804	Lily White	130	-	1	6	.01	.05	1	")			"
22													
23	4851	Almost Black	120	-	3	49	.02	.41	5	60)			12x10x9
24	4852	Espresso	"	-	4	71	.03	.59	7	")	1		"
25	4853	Bankers Gray	"	-	2	32	.02	.27	3	")			"
26	4856	Black Raspberry	"	-	2	37	.02	.31	4	")			"
27													
28	4901	Moonlights	225	-	14	263	.06	1.17	14	27		½	16x12x13
29	4902	Dawnlights	"	-	14	265	.06	1.18	14	"		½	"
30	4903	Sunlights	"	-	10	196	.05	.87	11	■		½	"
31													
32	5001	Ebony	130	-	1	4	.01	.03	1	36)			18x18x5
33	5002	Dark Tortoise	"	-	1	4	.01	.03	1	")	1		"
34	5003	Navy Blue	"	-	1	15	.01	.12	2	")			"
35	5004	Sea Green	"	-	2	25	.01	.20	2	")			"

Exhibit 7-1 (Cont'd.)

your company experiences its peak storage requirements in October (for Christmas sales), that month's inventory requirement would be used. Additionally, the Sales department should be requested to project sales five years into the future. The result of these calculations will be a total number of cases required for each item in inventory.

Next, case dimensions for the major volume items which cannot be bulk stored are reviewed and a standard pallet size is selected. It then becomes a simple matter to determine total pallet requirements by item. Small-volume items may share pallet space, be placed on decking in pallet racks or be stored in appropriate size bins.

The problem is then narrowed to selecting the appropriate type of rack or combination of racking systems to best suit your product mix and volumes. Listed below are the basic types of racks available on the market today:

(a) *Standard pallet racks.* These racks are available in a wide range of standard lengths, heights, widths and storage capacities. They usually hold one pallet in depth with adjustable cross-beams spaced to suit individual requirements. These units are usually placed back-to-back and are bolted together with "spacer bars" (to allow for pallet overhang). Select a rack with superior cross-beam locking devices. This is often the Achilles Heel of racks.

(b) *Drive-in racks.* Here, cross-beams are replaced with horizontal beams. These racks can be constructed to any in-depth dimension. They are particularly useful in storing large-volume items. The in-depth storage which they offer minimizes aisle requirements. But think small also. Drive-in racks can be used to store two pallets in depth as well as 24 or 36. In fact, used with a reach-type fork truck, you may store different items vertically. The "reach truck" works from the aisle and extends into the rack to obtain the second (rearmost) pallet without disturbing pallet loads above or below.

(c) *Cantilever racks.* While more costly than many other types of racks, cantilever systems offer post-free flexibility. They are the most practical solution in storing bulky items such as furniture, or randomly long items such as sheet steel, tubing and lumber.

(d) *Gravity feed rack systems.* Gravity pallet racks are the most expensive of the racking systems and offer the least flexibility. They are particularly advantageous when associated with automated picking systems.

The most successful application of the gravity feed principle has been in the less-than-full pallet load areas, such as in manual order selection lines. Typical applications may be seen in the cosmetics and pharmaceutical industries. Using adjacent conveyors and tote boxes, order fillers walk along the gravity feed racks, selecting individual items along with full carton lots.

Whichever racking system is selected, be sure to select materials handling equipment which is compatible with the other parts of the system; i.e., racks, pallets, aisle dimensions, floor loading capacities, overhead obstacles, delivery vehicles and dock areas.

(e) *High-rise storage systems.* Investment in storage systems over 30 feet high has gone from virtually zero 14 years ago to an estimated $130 million at present. These systems have been developed to provide users with greater lifting capacity to greater heights than conventional counterbalanced fork-lift trucks or narrow-aisle straddle trucks can deliver. High-rise systems offer significant savings in terms of land and building costs.

D. ORDER PICKING AND THE EFFECT OF PACKAGING

A major segment of warehouse expense is found in the picking of goods and delivery to the packaging or shipping area. Economy in picking is predicated on the location of goods so that economic and expeditious picking is possible. A corollary requirement is the listing of goods to be picked in location sequence to minimize travel.

The degree of success in the above will vary with the type of warehouse. A hardware house with thousands of items of various applications will offer a greater problem than a grocery warehouse where there is a substantial similarity of package.

In picking unitized items to be loaded directly into railroad cars or road vehicles, sequential picking does not become a factor since all loads picked are uniform and load distribution in the vehicle is the governing factor. In picking a large number of different items of varying sizes and in varying quantities, sequential picking is the main factor in speed and economy of delivery to the packaging area.

In a sequential operation, it is necessary to set up the shipping order in sequence of item location. For example: in a drag line operation in a shoe distribution warehouse, the sales order is set up in sequence of location along the line and then by styles and sizes. The order is attached to the cart and a number of cards placed in a holder, with the outer card readily visible. Each card bears the number of the area and the cards are in the same area sequence as the order. Each picker as he fills his part of the order places his number card at the rear with the next area number exposed.

A drug warehouse picks into baskets on a conveyor. The order again is sequenced by location number and the picker moves along, picking from bins or flow racks on both sides of the conveyor.

All are basically the same, but because of conditions peculiar to that particular operation, there are method variations.

Electronic information flow has speeded up order processing by automatically analyzing a customer's order and converting the information to the required sequence. This has contributed greatly to automatic picking.

It should be noted that at present there are very few completely automated warehouses with all items picked automatically and with automatic replenishment. In some fields this will always pose a problem. Commercial warehouses and metal warehouses with a variety of sizes, shapes and weights will probably never be potential applications.

At present most existing warehouses that have automatic picking move 70–80% of the volume through this path. This, however, is only 25–30% of the items in the warehouse. The balance are subject to other methods of picking, usually manual.

As you can appreciate, packaging is the key to automatic picking. The design of the package unit has a definite effect on the pattern of storage, picking and handling.

With the development of integrated distribution systems with automatic warehousing, containerization and special transport equipment, the impact on the warehousing operation becomes even greater.

Uniform packages are the basis of the unit load system. Packaging has had other impacts in the warehousing:

1. *Reduced damage*—The uniform package has allowed better and safer stacking. Package design has contributed to better protection of goods in the warehouse.
2. *Space utilization*—Package design has contributed to space utilization by the use of greater stack heights.
3. *Distribution expense*—Here, package design has produced stronger, more compact packages that have allowed better stowing of freight in vehicles. Unprofitable tare freight expense has also been drastically reduced.

E. MANAGEMENT PROBLEMS IN THE SMALL WAREHOUSE

The small warehouse will usually have all the problems of a larger facility, unfortunately, magnified out of proportion. Usually it is faced with existence in facilities not specifically adapted to the operation. As a result, management's major objective must be the minimization of the several problems. They are rarely eliminated.

The problems usually are:

1. *Limited space*—Poor column spacing, low ceiling.
2. *Unplanned growth*—This may result in several levels or buildings.
3. *Inadequate layout and construction*—Low floor capacity.
4. *Low percentage of unit load handling*—Goods may arrive in large quantities but are issued in small and varied orders.
5. *Small or no application of mechanized equipment*—Usually the physical facilities prevent mechanization.
6. *High labor costs*—Due to all of the above conditions.

The solution will be limited and will call for considerable brainwork by management. Emphasis must be placed on flexibility of facilities, equipment and labor. It may be necessary to improvise on standard equipment to arrive at an optimum operating condition.

The information flow to control the movement of goods must be kept at a minimum. Normally all that is required will be:

1. Receiving papers.
2. Inventory records.
3. Shipping papers.

A sophisticated information flow system would be expensive and unnecessary.

The following are two examples of what can be done with an inadequate facility. They cover the physical aspect which usually is the most expensive.

The first is concerned with the warehousing of 5-gallon, unpackaged cans, and cartons of 1-gallon cans. The plant's products were usually shipped in tank cars or trucks or in 55-gallon drums. Small package warehousing was a stepchild. The warehouse facility was an old processing building with three floors serviced by an elevator. Goods arrived via tractor-trailer train from the filling lines.

Normal procedure called for manual transfer of the cans or cartons from the carts to a pallet, movement by pallet jack to the elevator and transfer to the upper floors for storage one pallet high. Ground level was assigned to shipment accumulation and other service functions. Transfer to the shipping dock was by tractor-trailer train. When an order was received, a picker took a pallet or a pallet jack and picked the goods from the various stacks, moved the pallet to the elevator and manually transferred the goods to the waiting carts.

The solution called for an inclined, powered, two-way conveyor between floors and a gravity conveyor with power boosters making a loop on each floor. These conveyors serviced storage areas on either side. By checking the production schedule, the warehouse foreman could determine his space requirements and select storage locations. No item ever exceeded the capacity of the conveyor loop on each floor.

Goods were unloaded on the lower level to the powered conveyor and a stop placed on the conveyor on the storage floor. When unloading was completed, the materials handler moved to the storage floor and stacked the material in storage. Greater space utilization was possible. On picking, the warehouseman placed the material on the conveyor, which transferred it to the lower floor where it could run out on accumulation spurs. The resulting labor savings paid for the installation in less than a year and service was better.

Another example concerns an operation in which the main product was cartons of 24 1-quart cans and a supplementary movement of cartons of six 1-gallon cans. The problem in this warehouse was rapid loading of road equipment. There were only three truck spots; there were favored customers who had preference over other waiting trailers.

The warehouse was separated from the truck dock by another building. Cartons moved on a two-level conveyor to the dock. All storage was floored at the conveyors in the warehouse. Imbalance of flow from the warehouse to the dock was inevitable since the warehouse conveyors also serviced the canning machine on new production to be stored.

Some unused space between the other building and the truck dock was found. By erecting storage conveyors, it was possible to store about a trailer load of cartons behind each truck spot. This surge location had two advantages:

1. *Trailer loads*—A trailer load of mixed items could be accumulated and quickly loaded.
2. *High-volume items*—This product had most of its large volumes in a very small

number of items. The storage conveyors could be loaded with large lots of these and the smaller quantities sent by conveyor from the warehouse.

F. TRUCK DOCKS

1. *Dock location.* The most common procedure is to locate shipping and receiving docks adjacent to one another. This arrangement facilitates vehicle check-in and control and also minimizes supervisory requirements.

The second most popular method is to separate shipping and receiving. The rationale here is to remove congestion, improve product flow and security.

2. *Truck berth width.* Adequate berth widths improve operating efficiency and reduce damage to both trucks and dock areas. Spaces should be no less than 12 feet wide with 14 feet the recommended width. Over-the-road trailers now measure 96 inches in width. Both the trucking industry and the ATA are lobbying for an increase to 102 inches.

3. *Maneuvering and loading areas.* Exhibit 7-2 shows the minimum dimensions recommended for maneuvering and loading areas. Note that the loading area is oriented so that the prevailing wind runs across and not into the door openings.

The trend towards longer trailers strongly suggests that provision be made for overall vehicle lengths of 60 feet. In addition to the area shown in Exhibit 7-2, space should be planned for marshalling and temporary storage.

G. WASTE REMOVAL

1. *Collection systems.* Warehousing waste removal normally involves a two-step process; i.e., collection and processing.

Collection equipment ranges from the common trash barrel to modern compaction units. Wirebound crates and metal containers are frequently used also. These may be placed on pallets, skids or they may be castered. Waste is ordinarily then removed to larger containers, either closed or open, on the dock or elsewhere outside the warehouse. These containers are used to accumulate large quantities for pickup at convenient intervals. They usually are of three types: self-contained trailers, large roll-off units or smaller containers which are hoisted onto trucks. Loaded containers may be connected to on-site compactors or they may be hauled to incineration stations, transfer sites or landfill locations.

2. *Compaction.* Compaction's major virtue is, that in reducing waste volume, the number of hauls away from the warehouse is minimized. This results in the lowest possible transport cost. Their crushing action is a desirable security feature to sensitive operations such as drug and cosmetics distribution.

Compactors can handle just about anything that a warehouse can generate in the way of waste. Heavy-duty units easily accommodate wood pallets and steel drums.

Generally, the maximum size of the trash determines the size compactor to purchase. Price differentials between large and small units are relatively small.

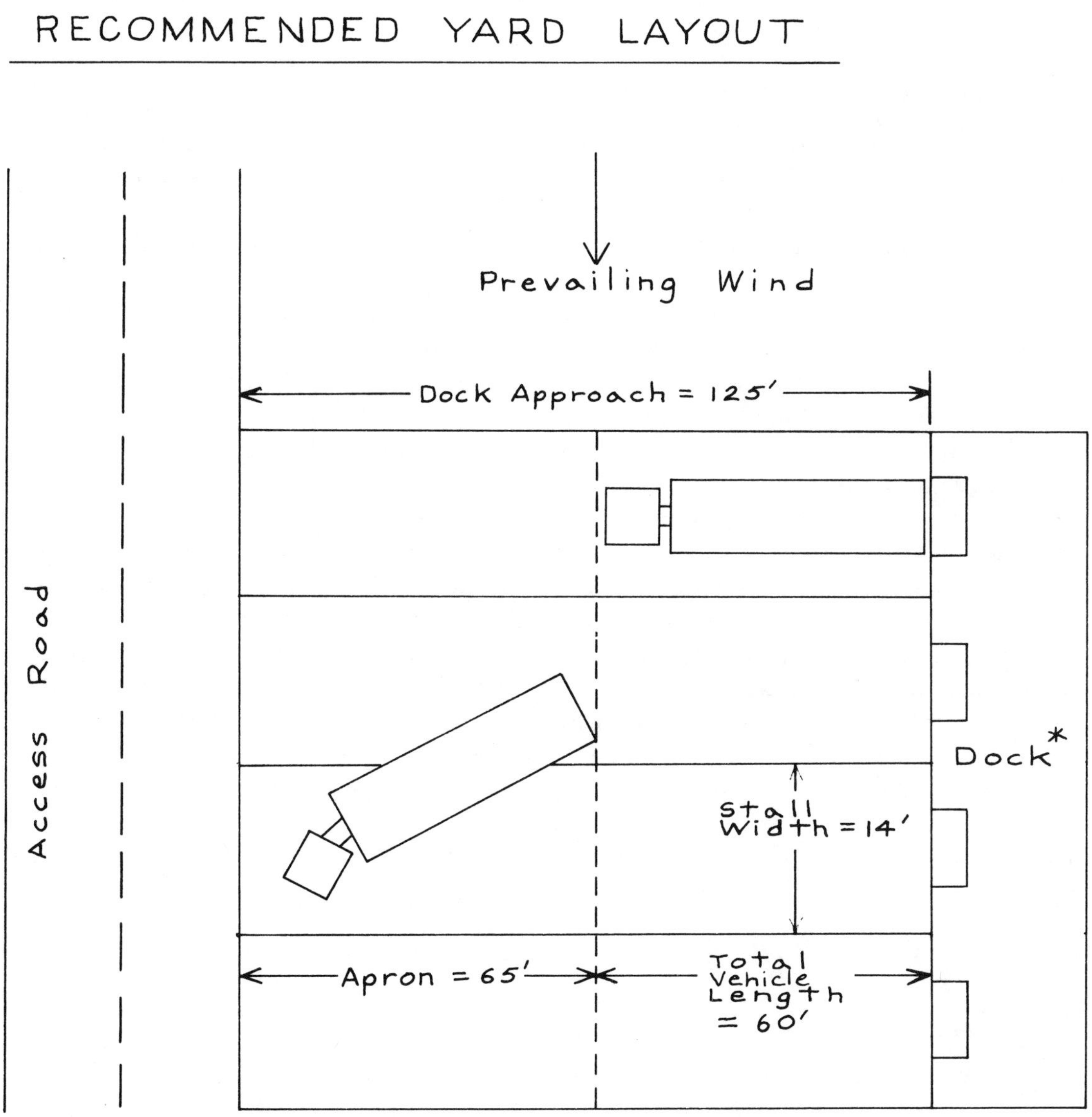

*Depth of dock depends on factors such as: shipping and receiving procedures and volumes, accumulation requirements, type of materials handling equipment being used and pallet dimensions.

Exhibit 7-2

3. *Baling.* Balers have been around for a long time and still serve well where salvageable materials such as corrugated cartons and paper must be handled. Balers accomplish the same purpose as compactors, at low cost for low-volume users.

H. AUTOMATED WAREHOUSING

There has been much published about automated warehousing; however, there are very few completely automated warehouses. There are close approximations in materials movement. You will find the petroleum industry as close to it as anyone. After all, a refinery is nothing but another unit in the movement of material from the well to the gas tank of your car.

The industry is at a point where computers can tell the best pattern of processing a specific crude oil and the best rate to meet a specific sales requirement. In fact, the computer can predict the ideal sales requirement for a particular type of crude. Based on that, it can run the refinery with the operator as a watchman.

We do not have this in warehousing. Rather, we do have degrees of automation or automatic warehousing.

With large movements of identical items, it is possible to automatically pick goods and transport them to an accumulation or loading area. You will find in all sophisticated samples of automatic warehousing that about 70–80% of the volume can move in that manner. This will be about 30% of the items. The balance will fall into the current concept of warehousing: selection of a number of items manually (perhaps supplemented by mechanical equipment) and transporting them to packing, and then accumulation and loading. In a large-volume, unitized movement, automated physical movement is now not practical. The fork truck with attachments such as grabs, extensions and pushers, is still the most economic means of picking, transport and stowing of freight.

As you automate, you become concerned with several physical categories which move from manual to automatic control in varying degrees:

1. Materials handling.
2. Stock selection.
3. Transportation.
4. Packaging.

Corollary to the above are the service functions:

1. Inventory control.
2. Paperwork processing.
3. Personnel.

Automation in warehousing is defined as a continuous and integrated system which uses electronic and/or other equipment to regulate and coordinate warehouse operations. It is concerned with:

1. Analysis of basic data.
2. Resulting calculations.
3. Regulating the material volume and flow according to the results of the calculations.

The basic steps on automating a warehousing system are based on one question, "To what extent shall we automate?"

What operations should be automated?

1. Order processing.
2. Inventory control.
3. Warehousing.
4. Materials receipt and delivery.
5. Picking, packing, accumulation and loading.
6. Paperwork.

What degree is practical? What is the degree of automation? What are the costs?

1. Material.
2. Labor.
3. Overhead.
4. Capital expenditure—automation is expensive.

What are the benefits?

1. Savings.
2. Service.
3. Inventory levels.

What problems will appear?

1. Disposal of personnel.
2. Retraining.
3. Unions.

All the above will have an impact on an intelligent evaluation of the degree of automation.

To complete the evaluation, the desired system must be analyzed in terms of basic warehouse requirements:

1. Space.
2. Facilities.
3. Equipment.
4. Paperwork.

5. Costs.
6. Savings.

Since automation in most degrees is based on E.D.P., the following will illustrate an application to a distribution operation. It concerns a parts replacement program for an appliance manufacturer.

1. Distribution transmittal of required data by wire to service center.
2. Converted at center to punched cards.
3. Fed into computing system:
   a. Customer account record scanned.
   b. Inventory updated and checked.
   c. Availability and location determined.
   d. Pricing.
   e. Order cards produced.
4. Order cards processed for order.
5. Picking tickets produced.
6. Extension of process will produce shipping label, postage requirements, bill of lading, inventory adjustment, routing and invoicing.

This could possibly be extended to robot picking. As yet, this has not been developed. At present the main claim is reduction in the order cycle from two weeks to about two days.

CHECKLIST TO TEST MATERIALS HANDLING EFFICIENCY

1. Material and Product
   A. Material too large for handling equipment.
   B. Material does not nest or stack.
   C. Material cannot be piled to form unit load.
   D. Travel paths long or circuitous.
   E. No definite pickup or delivery points.
   F. Backtracking between operations.
   G. Material moves between floors and buildings.
   H. Bulky materials fabricated considerable distance from point of use.
2. Ordering and Operations
   A. Orders frequently partialed.
   B. Small orders moved independently.
   C. No move authorization on material.
   D. Equipment failure causes frequent backup of materials on dock and vehicles in yard.
   E. Extra handling due to lack of scheduling.
   F. Excess received goods or damaged materials not quickly processed.
   G. Transfers from one handling device to another.

3. Handling Organization and Systems

   A. Material handlers searching for items.
   B. Fork-lift trucks moving empty.
   C. Equipment operators waiting for instructions to move materials.
   D. Call or dispatch system inadequate.
   E. Materials handlers not zoned for efficient operations.
   F. Freight car demurrage expense high.
   G. Waste materials not removed quickly or on a scheduled basis.
   H. Personnel not available to receive material.

4. Handling Equipment Use

   A. Equipment not available because of repairs or because being used by others.
   B. Equipment infrequently used.
   C. Equipment has insufficient carrying capacity.
   D. Hand trucks used for many materials movements.
   E. Materials and storage equipment being damaged too frequently.

5. Building Hindrances

   A. Rough floors cause slow travel for careful handling.
   B. Slow-speed elevators.
   C. Floor level loading or unloading of freight cars and trucks not possible.
   D. Doorways too narrow or too low for full loads.
   E. Low floor load capacity.
   F. Dock levelers not used.

6. Use of Space

   A. Equipment moving too slow because of very narrow aisles.
   B. Excessive aisle width.
   C. Vertical space not used for storage.
   D. No location system or assigned space for materials.
   E. Aisles being used for storage.

8

Making Value Engineering Work for You in Reducing Distribution Costs

In order to maintain our competitive position in domestic and international markets, a new approach is needed to meet customer demands. It is necessary to deliver a product at the lowest possible cost to perform an essential function. This means giving the customer more for his money. To do this, it is mandatory that we control value. This requires a new approach to delivering values. The new approach is called Value Control.

This chapter shows you how to introduce and implement a value control program in your distribution operations. Detailed coverage is given to the following important factors:

- A new way to present the "value" and profit contribution which Physical Distribution can make to your company.
- A systematic, step-by-step procedure to use in starting a value control program.
- Several examples which show how distribution managers in different businesses made value control work for them in achieving significant dollar savings.

A. VALUE CONTROL

Value control was born through necessity, the necessity to get costs down. It is a technique that supplements conventional cost reduction programs. Simply defined, value control is an organized approach to giving the customer a product which is more valuable because it will perform an essential function as well or better than before, at lower cost. Its aim is to give the customer more for his money. The physical distribution organization, which accounts for 10% to 40% of cost of sales in most companies, can play a significant role in this effort.

1. *What is value?* To understand how value control works, it is first necessary to understand value. Value is hard to define because each individual has his own personal concept of it. Like the concept of beauty, the concept of value varies.

Just what is value—how is it determined? Let's approach it in the following way. What is the value of a small bolt in an airplane wing if the failure of that bolt causes the wing to fail, resulting in the death of passengers and crew? Our first reaction is probably that this is a pretty important bolt and no expense should be spared to build it "right." Certainly, no one would suggest using an inferior part for such an important job.

However, could the value of the bolt ever be equal to the price of the airplane or the value of the lives of the passengers and crew? Certainly not. To make a proper approach to determining the value of the bolt, one must recognize that value has no relationship to either the cost of producing a part or the consequences in the event of the failure of the part. The actual value of a part or device can be determined in only one way, by comparison with other ways of performing the function.

Value represents the lowest price one must pay to provide a reliable function or service. To determine the value of the bolt, one must compare the various ways of accomplishing the function of holding the wing on the airplane. The one that represents the lowest cost, consistently, with reliability, determines the value of the bolt. There are four kinds of value: esteem, exchange, cost and use. We are talking about the *use* value of an item.

Most of us do well at determining use value of items. For example, when you see a price tag of $100 on a pair of shoes it probably jars you a little. The reason is simple. You know from experience that you can buy a good, functional pair of shoes for $30 or so. The value control approach demands that we develop a **sense of value** concerning the things we do on the job.

This isn't always easy because value is an ever changing thing. What is good value today is not necessarily good value tomorrow. Value is determined by the technological and productive capabilities existing at a point in time. Because of this, value analysis is an ever continuing need. The technological and productive capabilities are continually advancing. For example, one could hardly have been expected to use high storage racks to reduce total building costs before the development of the modern fork-lift truck.

The main theme of value control is: Spend company dollars as carefully as if they were your own!

2. *Establish possible value control attitudes.* Knowing more about value should cause the term value control to take on new meaning. Here is the basic philosophy that makes up the value control approach:

(a) A better job can be done. Everyone can do a better job in contributing to the value of a company's product. If you don't have the best design, the best material, the best methods, the best vendor, you can and should do a better job. Now, "the best" does not mean the most complicated and costly, but the simplest way consistent with reliability that will perform the function.

(b) Value control requires a team effort of all persons involved with product design, production, procurement and distribution, pulling together to accomplish the best results.

B. TECHNIQUES OF VALUE CONTROL

Value control is a fresh approach to the old problem of keeping costs down. It does not eliminate nor reduce the need for conventional cost reduction through methods improvement, effective purchasing, inventory controls and other established methods of saving money. The value control approach requires a constantly questioning attitude. The following are the important techniques of value control:

1. *Determine the function:* Clearly determine the required function. Exactly what does the product do; what does each assembly on it do; what is the function of each part?

2. *Evaluate the function:* Evaluate by comparison. What does it cost to accomplish each of these functions? Might they be accomplished by alternate ways and what would the alternate cost be? Compare the costs for accomplishing somewhat similar functions by diverse means.

3. *Creativity:* Use creative thinking. What else will do the job? Allow the imagination to soar. This mental activity could be described in terms of a fisherman who progressively casts from a boat in different directions, covering the entire area in a systematic manner and avoiding any pre-conceived notions based on faulty information as to where the fish are. These creative studies identify the areas where more intense value control work can be done.

4. *Analyze:* Blast, then refine. Blast the item under study down to one which will not accomplish the total function but will be extremely simple and have low cost. Then progressively refine this basic product by adding required increments of function. This technique usually produces designs that are more functional, more reliable, more simple and have much less cost.

5. *Use specialty products and services:* Use available functional specialty products and specialty facilities. The United States is filled with companies whose success is built upon accomplishing some specialized job extremely well. These companies frequently make needed products in large volume on highly specialized equipment. By the incorporation of such parts into the design of materials handling equipment, the functions they provide are often accomplished at much less cost than normal design and in-plant fabrication work.

6. *Gather facts:* Get all the information from the best available sources. Experience shows that although the "best" source can usually be identified, answers are often accepted from others. For example, a specialty vendor is the best source of information on his capabilities, not a buyer who has bought items from him.

7. *Human relations:* Watch the human relations angles. Obtaining better value is primarily an activity in human relations. Good value control work will show up many needed changes, some of which should have been made sooner. Such obvious savings can be extremely belittling and embarrassing to the individual. For this reason, the correct handling of human relations must have top attention. The emphasis must not be placed upon who or whom is responsible for a part, piece of equipment, pay-out of a system the way that it is, but upon how its value can be improved.

8. *Know your costs:* Obtain accurate and total costs. Such figures are often hard to get; however, it pays large dividends when the difficulties are overcome and meaningful estimates and costs are obtained.

9. *Plan your approach:* Follow a carefully planned approach. To help you in this respect, the step-by-step approach to value control called the Value Engineering Job Plan is presented next.

This job plan, used as a guide, will help you avoid the trap of holding out for a perfect solution.

C. THE VALUE ENGINEERING JOB PLAN

1. *Information phase.*

Secure all the pertinent facts. Get actual samples of parts or products under consideration. Obtain cost figures, quantities used, vendors, specifications, distribution and manufacturing methods information.

2. *Speculative phase.*

(a) Generate every possible solution to the problem.

(b) Consult others who may help.

(c) Systematically explore various materials, machine processes, rearrangement of parts, etc.

(d) Use imagination freely. Encourage others to make use of their imagination.

(e) Record every suggestion that seems remotely possible.

(f) Establish two-man teams for the creation of additional ideas.

3. *Analytical phase.*

After all of your ideas have been written down, you are ready to start, "Engineering them down to earth." Remember the following:

(a) In this phase, you should not try to eliminate ideas, rather try to make the ideas work.

(b) Take the positive approach. Do not allow negative thinking or prejudice to enter the picture.

(c) Consider all the ideas that will perform the function, regardless of existing specifications, interchangeability, etc.

(d) Place emphasis upon the ideas in proportion to their value and the probability of accomplishment.

(e) Select at least two of the "best" ideas. List both the good and bad points. Overcome objections which are not valid.

(f) Select the top specialists in the company for their input when pertinent.

4. *Planning and preparing for action phase.*

In the analytical phase, ideas previously generated were screened, and those showing the greatest potential were selected for further consideration.

Now, in this phase, you are ready to take the best ideas and gather additional support information that you will need to develop them into a sound, useable suggestion.

You will need specific and accurate data to fully support your final recommendations. Specialists within your company and specialty suppliers outside your company can be helpful in this regard.

Stick to each promising suggestion. Thrash it out and reach definite, tangible, useable conclusions.

5. *Presentation phase.*

To continue your planning and preparation for the later "action phase," you must put your recommendations and findings into a saleable package. If properly prepared, your report will demand the necessary attention to turn your efforts into real accomplishments. It will help to install your ideas.

For your report to communicate, you must consider those who will see it. Tailor-make the report to ensure that it will gain proper attention. Those who read your report may not be familiar with the subject. The report must tell them all they need to know about your value improvement idea. The before and after must be clearly explained. Review the before briefly. Justify the after. Supplement your report with graphic materials.

A written report alone may not be sufficient to sell your idea. If possible, support your proposal with a personal presentation. Plan it carefully. Rehearse it sufficiently to assure a smooth delivery and accurate timing.

6. *Action phase.*

After your ideas have been accepted, action must be taken to assure that your proposal becomes a reality. As the idea moves toward installation, it is possible that opposition will increase in some areas. Do not be discouraged by opposition. Remember, your job begins, not ends, with roadblocks.

A basic advantage of value control is that it produces measurable results. The profit shown is the "proof of the pudding" stage of the value engineering effort.

To summarize, the application of value control requires that, for each item of cost we ask:

- What is it?
- What does it do?
- What does it cost?
- What is it worth?
- What else will do the job?
- What does that cost?

In answering these six questions, we have the opportunity to draw upon our own experience and the experience of others, and to use whatever other methods of investigation and search are available.

Remember that cheapness does not always reflect less total cost or less long-run cost. You may find that a cheaper part may create problems later in cost of maintenance. Consequently, value control goes beyond conventional cost reduction thinking, in considering not only present costs but also ultimate costs and their effect upon the downstream value of the product. Value control thinking might even result in a decision to increase the cost of an item, when this is justified, to increase the ultimate value of the product to the customer.

D. EXAMPLES OF COST SAVINGS THROUGH VALUE ANALYSIS

The way you present ideas to management can make or break a project. Exhibit 8-1 shows a typical "cost comparison" display. Don't forget to calculate implementation costs; sometimes these can add up to more than the anticipated savings. Several examples of cost savings through value analysis are shown below:

1. A chemical company, shipping a granulated product in corrugated containers, first used an inexpensive resin glue to seal cartons. Subsequently, it was found that the leading edge of the flaps was getting caught on the flooring of trucks and railroad cars, resulting in spillage and customer complaints. Unreinforced tape was then added to gluing so that carton flaps would not grab. Customer complaints now revealed that when stacked five cartons high, the tape split under the pressure. Metal strapping was added to the gluing and taping operations. Customer complaints ceased. The job was finally done satisfactorily. But at what cost?

A value analysis team selected carton sealing as an area of investigation. The result of their investigation showed that a "hot melt" carton sealing operation, replacing the other methods, resulted in annual savings of $75,000.

2. A cigar manufacturer found that its packaging costs could be reduced $45,000 per year by replacing nails with automatic tape dispensers in sealing its cigar boxes.

3. One of the leading textbook manufacturers in the country found that using disposable pallets over conventional wood pallets saved $12,500 annually.

4. A distributor of children's books saved $9,500 per year through a value analysis project which demonstrated that shrink-wrapping of customer shipments onto corrugated trays was superior to stuffing cartons with scrap paper. In addition to the cost

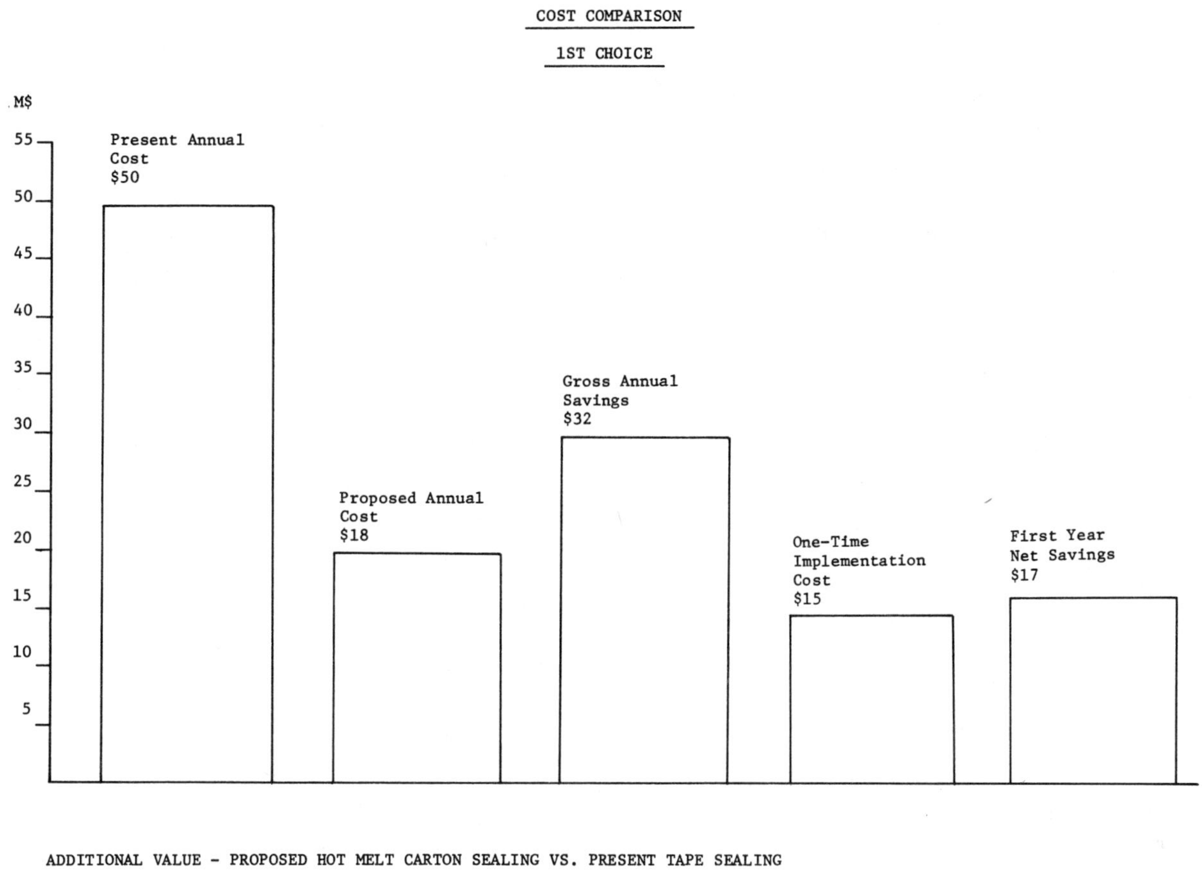

Exhibit 8-1

savings, shrink-wrapping resulted in a significant reduction in product damage and customer complaints.

5. One value analysis team in a plastics company saved $25,000 per year in evaluating carton sealing methods. Material was packaged in corrugated containers under pressure to increase density. Reinforced tape was used for sealing. Value analysis demonstrated that pneumatic strapping was faster, surer and less expensive.

6. After achieving its first successful investigation as outlined in 5 above, the value analysis team went on to examine its truck delivery operations. They found that in some areas, obtaining their own leased trucks would be more economical than using common carriers. Annual savings amounted to $125,000.

7. Through value analysis, a distributor of chemical supplies found that it could save $35,000 per year in its order filling operations. Individual fork trucks, which had been used for carrying less than pallet load orders, were replaced with tractors towing three trailers each. Fewer trips to the shipping dock were needed and less traveling was required in the order selection process.

8. An electronics instrument manufacturer significantly reduced product damage and also saved $15,000 per year by conducting a value analysis examination of its packaging and shipping operations. Orders were shipped in large corrugated cartons filled with excelsior. This was found to be slow, and it also resulted in customers returning units for recalibration. These problems were eliminated by using a sleeve-wrap bundling system. Here, unit cartons of product are placed on a heavy-duty corrugated tray with polyethylene shrink film. Each individual carton acts as a cushion around the product itself. The package is completed by placing the unitized bundle into a corrugated sleeve with an end cap and strapping it together with non-metallic strapping.

9

Effectively Using and Maintaining Lift Trucks

This chapter presents, in detail, a five-part, systematic approach to getting the greatest dollar return on your fork-truck investment. First in importance, particularly in view of OSHA standards, is an operator program designed for safe and efficient lift-truck use. Part II is an equipment maintenance program. Some of the more popular attachments which can be added to your lift trucks to make them more efficient is covered in part III. Part IV involves the proper selection of standard pallet racks to assure proper functioning of your lift trucks. Finally, part V sets forth a successful method to be used to determine when your equipment has given the best that it had to give and exactly when it should be traded.

I. OPERATING TECHNIQUES

Your lift truck increases operating efficiency, enables you to produce on schedule, cuts handling costs and pays dividends by upgrading labor to more productive work. But fork trucks tend to be abused and neglected, and this can result in inefficiencies and excessive costs.

> *CASE IN POINT:* A building materials supplier having a fleet of 16 fork trucks, found that 25% of them were regularly in the maintenance shop for major repairs. Maintenance costs were running $800 per week. An investigation showed that night-shift operators were "racing" the equipment and engaging in other hazardous practices. After management took corrective action and these activities ceased, the

company found that it could operate with two less fork trucks and operators. Maintenance costs decreased to $150 per week.

All equipment operators should be trained in efficient and safe vehicle operation. Whether your training program consists of formally scheduled classes or a breaking-in-the-man routine, the new operator must prove conscientious about the A-B-C's of his job. From the maintenance viewpoint he should:

A. Have acquired correct operating habits and handling skills.
B. Have a knowledge of the functions and limitations of his truck.
C. Be able and conscientious in its routine inspection.
D. Recognize characteristics that are breakdown danger signals.
E. Be conscientious in reporting damage and/or breakdown indications.

Too often the careless and slipshod habits acquired in passenger-car driving are permitted to be carried over into industrial truck driving.

Training Program

Here are some pointers on what might be termed "preventive maintenance driving."

Starting: Close the choke button as soon as possible after a cold engine is started. Excessive choking will cause unburned gasoline to leak past the pistons and dilute the lubricating oil. In cold weather, always allow engine oil in hydraulic systems to warm up and circulate freely before operating. If the unit does not start in one or two minutes, the operator should report it to his foreman or other authorized personnel. He should not try to start it himself.

Shifting: In operating straight clutch drive machines, the clutch pedal should always be pushed to the floorboard before attempting to shift gears. This minimizes undue clashing of gears.

Riding the clutch is an all too common habit. This causes loss of tension on the spring and allows the clutch to slip, resulting in rapid wear. It can also damage the throw-out bearing by melting the grease through friction heat.

For engine protection and safety, always shift to a lower gear before descending an incline that requires excessive use of brakes. To make this shift, release clutch and shift to neutral, then re-engage clutch and adjust speed to permit shifting without clashing, finally declutch quickly and shift to the lower gear.

When going up an incline, shift to the lower gear before the engine starts to labor.

To reverse direction of travel come to a complete stop, then shift to reverse.

Traveling: Operators should not pump the foot accelerator while traveling. This brings about worn cylinders and cylinder walls, besides consuming excessive fuel.

Trucks should never be left unattended with their motors running. Gasoline engines wear most rapidly when standing at idle speed.

High-lift trucks should not be used to carry loads over great distances (that is, more than 300 feet). There are more economical methods and equipment for extensive horizontal movement.

The operator should have the authority to refuse to drive his truck over floors littered with chips, turnings and other tire-destroying scrap. Oil and grease on floors are also to be avoided since they deteriorate rubber tires. Railroad tracks should be crossed diagonally rather than at right angles, particularly when the truck is loaded.

Collisions: While safety considerations are the prime reason for avoiding collisions, they always result in truck damage. However slight the damage, all collisions should be reported so that the cause may be determined and eliminated. This is also preventive maintenance.

Loading: Responsibility for proper loading should rest with the operator. He should have the authority to refuse to move loads that may damage the truck. Overloading can result in broken lift chains and cables, can damage the hydraulic system, packing and cylinders and can overstrain engines and batteries.

The operator should know that, although a fork truck has a certain rated capacity, it cannot always handle loads of that weight. Fork trucks are rated on an inch-pound capacity basis, as measured by weight of load and its length from load center to a point on the truck specified by the manufacturer. This may be the heels of forks or the front axle.

Adding or removing attachments or accessories, such as a load grab, extension forks or a boom, affects the capacity rating of your truck. Rated capacities should be specifically discounted by the operator when he is traveling over uneven surfaces, down grades or when transporting overhanging loads. Warning systems that indicate overloading are available.

Stacking: If possible, the load should be raised as the truck is nearing the stack while coasting into position. This reduces strain on moving parts. When raising or lowering loads while standing still, the truck should not be left in gear with the clutch depressed. Excessive engine speed will not increase the speed of the hoisting mechanism and will only result in excessive wear. A good operator will learn to judge the proper engine speed by "feel."

Stopping: Speed should be reduced gradually. Sudden stops are severe on the braking mechanism and a sharp halt is liable to cause the load to tilt forward and spill.

No one should operate a truck other than the person to whom it is assigned. As much as possible, each man should work permanently with the same machine. This is important when the driver is under a production incentive system, where he may be inclined to overlook preventive maintenance regulations in favor of continued operation.

In case of apparent mechanical difficulty, it is the operator's responsibility to stop the machine and report the trouble immediately. He should not attempt to make repairs or adjustments himself, nor should he attempt to move a disabled truck except under the direction of a maintenance mechanic.

Here are danger signs which, if caught in time, will prevent major repair jobs:

1. Lack of power in truck, lifting mechanism or accessory attachments.
2. Faulty lubrication.
3. Worn or damaged tires.
4. Poor brakes, difficult steering.
5. Missing parts, broken or bent rods, etc.

Additional Pointers

Fire prevention on lift trucks is also preventive maintenance, and your operator should be instructed to avoid unnecessary fire hazards. He should refuel only in an outside location with the engine off. The truck should stand 15 minutes if excessive gas spillage has occurred. Flammable liquids and acids should be transported only in approved containers. Drivers should see that fire extinguishers are periodically inspected and charged, and above all make certain that their extinguishers are of the correct type.

Inspection frequency, and items to be included in the various inspections, will differ from firm to firm. Some of the factors determining inspection frequency are: surface over which the trucks operate; size and type of loads; atmospheric conditions, including temperature and dust; whether the service cycle consists of long or short hauls; amount of jockeying; actual hours under load and whether the truck is assigned to one or several operators.

Maintenance on equipment driven in dust-laden atmospheres requires special attention. First and foremost is the importance of greater frequency—sometimes as often as every 15 to 20 hours. Gas trucks should be equipped with a pre-air cleaner to filter out the heavier particles before they can get into the oil bath cleaner.

There are some additional requirements for operating in cold storage spaces. Special hydraulic oil and low-temperature lubricants should be used. All nuts and bolts should be cadmium plated. On lift and tilt cylinders, rods should be of stainless steel. Since contacts and switches vary from manufacturer's make and model, it's sometimes advisable to put on heavier return springs.

On electric trucks required to operate continuously in extremely cold temperatures, it is wise to enclose contactors and controllers in much the same way as on a spark-enclosed machine. Heating equipment is then installed around the master directional and limit switches.

A road test, under capacity load, wherein the truck and lift-tilt mechanisms are put through their paces by a competent mechanic—with this check, coupled with a thorough visual examination of the upper and under sides for apparent wear, leaks, cracks and damage—is the most efficient method for determining which items demand immediate and thorough attention. Finally this good rule of thumb: "If you can see it, inspect it."

Operator's 12-Point Checklist

These 12 points should be run through on every lift truck at the start of each shift:

- *Crankcase oil:* It should be on or slightly below the "full" mark of the "dip" stick—never above.
- *Radiator:* If it is found necessary to add water in any significant quantity, this may be an indication of leaks or overheating. The most common cause of overheating is plugged radiator cores or fins due to accumulations of dust and lint or other materials. Blowing out periodically with compressed air will keep the cores clean. A loose fan belt will also cause overheating.

• *Filling of fuel tank:* Use clean gasoline from clean containers. When stored outside in steel drums, there may be a great deal of condensation due to atmospheric changes. If there is a suspicion that water is present in the gasoline, strain through a chamois-covered funnel. The chamois will remove both the water and foreign matter.

• *Air cleaner:* Since the oil bath serves to clean air before it enters the carburetor, importance of this cleaner's proper maintenance cannot be overemphasized. A check of its cup is required every eight to 16 hours. In extremely dirty operations, it should be cleaned at least twice a day. The mesh screen will gradually accumulate dirt each day. And once every 40 to 50 hours, the complete cleaner should be removed and washed.

While this latter task may be the responsibility of the maintenance mechanic, the importance of a daily check should be made clear to the truck operator.

• *Tires:* Pneumatic tires should be checked for specified pressure. Both pneumatic and solid-cushion tires should be examined for abnormally bad breaks, cuts and gouges. Foreign matter should be removed.

• *Gages* should be checked with the engine idling. The ammeter should show a reading according to the output of the generator, especially when the engine is accelerating. The oil gage should show manufacturer-specified pressure.

• *Brakes:* Pedal should resist pressure when approximately one-third of the way down.

• *Steering:* Operator should report more than 4 inches of play in the steering wheel.

• *Lift-tilt mechanism:* Work the hydraulic control levers and check for smooth operation of lift and tilt devices.

• On *gasoline-powered trucks* with conventional clutch drive, give the clutch throw-out bearing grease cup one to two complete turns after every eight hours of operation. Make sure the cup is kept filled with manufacturer-recommended lubricant.

• On *battery-powered trucks,* the battery should be checked to make certain it is fully charged. When fully charged or when changed between shifts, the new battery should always read 1,275 or higher.

• *Finally:* Check horn, lights and other special accessories.

II. PERIODIC CHECKUPS

Service life of your lift trucks can be increased immeasurably by a planned program of systematic inspections and lubrication.

Once established, your preventive maintenance program will keep on working for you through the years, long after the initial costs have been amortized. Dividends will accrue, too, in the form of lower service budgets, reduced truck down-time and smoother-running plant operations.

In addition to the regular daily inspection conducted by the operator, proper servicing of a lift truck necessitates three other regularly scheduled inspections: (1) weekly; (2) monthly and (3) semi-annually.

Recommendations of many truck manufacturers for the weekly (50-hour) inspection are as follows:

Weekly Inspection

• *Washing and cleaning:* Machines should be washed and cleaned completely. Where grease is present, a steam hose can be used to speed up work. Blow out radiator fins with compressed air. When the truck has been used in a dusty environment, remove covers from motors, controllers, contactors and limit switches. And after bleeding the plant air line of moisture, gently blow out dust that may have worked into these parts.

• *Lubrication:* Always follow manufacturer's recommendations when lubing trucks. In general, this operation consists of an "alemiting" steering connection and other bronze bushings. Some firms make blowups of manufacturers' lubricating charts for hanging on the wall of the Maintenance department to serve as a reminder and guide.

Keep distributor cup filled with grease. Use no more than one drop of oil on the distributor cam pad. Oil generator and starter motors lightly, one or two drops in each cup.

Check oil and/or grease level in drive axle, steering axle and steering gear box. Change engine oil and oil-filter cartridge. The usual viscosity recommendations are:

SAE 10—below 0°F.
SAE 20—0° to 32°F.
SAE 30—above 32°F.

• *Air cleaner:* Remove air-cleaner body and cup and wash in kerosene or other cleaning solvent. Dry with compressed air and refill to oil level, or slightly below if the truck is used on ramps. Also clean fuel screens on fuel pump, carburetor and gas tank.

• *Steering gear:* Check for security at mounting. There should be no more than 4 inches of play in the steering wheel. Examine rods and lever. If rods are bent, straighten them.

• *Brakes:* Test service (foot) brake for stopping under full load. Test parking brake for holding full load on the steepest incline on which the truck is used. Foot brake should take hold upon being depressed at 2 inches (one-third of the way down). Hand brake should hold at three to five notches or "clicks" on the ratchets. If it doesn't, brakes must be adjusted.

• *Battery:* Be careful with batteries. Overwatering will weaken the electrolyte. The exterior of the battery and battery case should be washed carefully if there is evidence of corrosion, and electrolyte level should be kept 1/4 inch above the plates.

• *Hydraulic system:* With forks down, check oil level in the hydraulic tank. If foreign matter is observed, drain immediately and refill with clean oil. Inspect both rigid and flexible lines for leakage. Check pump and valves for leakage.

• *Lifting mechanism:* Lubricate according to manufacturer's recommendations. Weekly maintenance should include lubrication of hoisting-piston head-sprockets and upright supports. Inner-slides should be lubricated with graphite paste. Chains should be examined, cleaned and wiped with oil (SAE 10).

Hoist and tilt cylinders may show leakage at the stuffing boxes on new trucks soon after they have been put into operation. This is due to compression resulting from wax impregnation of the packing rings. Should this occur, add additional rings and pull up snug, but not too snug or you will cause premature wear.

- *Forks:* Check forks to ascertain whether twisting or distortion has occurred.
- *Radiator:* Check radiator cap and replace gasket if lost. Blow out cores if plugged with lint or dirt, and make certain the overflow pipe is clear. Examine the water. If oil of any quantity is present, determine the source. Examine for general leakage.

Much of the foregoing can be applied to both gas and electric trucks. Here, now, are some items that apply only to electric-powered trucks:

- *Battery:* Keep battery clean and dry. And give it an equalizing charge according to maker's specs.
- *Seat brake:* Where a seat interlock is used, check adjustment. Brakes should not drag when seat is down, but should lock firmly when seat is up.
- *Drive motor:* Check motor insulation and terminal studs. The latter should be tight and coated with insulating paint. Tighten drive motor belts.

Monthly Inspection

There should be great care in the performance of the monthly (300-hour) inspection, since it covers inspection of parts on which wear may not be immediately apparent. And all of the items previously listed in the weekly inspection should be repeated at this time.

In addition, tighten cylinder head nuts with a torque wrench—not by feel. The recommended sequence for tightening should be used. Examine edges of head gasket for leaks. Check heat control valve for proper functioning and examine the exhaust manifold for leaks. Tappet valve adjustment and ignition timing should be checked during the monthly inspection. Also check the brake fluid level.

Look over all nuts and bolts. If there is evidence of corrosion, eliminate the cause. Remove corroded nuts and bolts, clean and cover lightly with grease.

- *Distributor:* Clean grease from outside cover. Inspect rotor, cap and condenser. Inspect and adjust the contact points. "Blue" contact points are caused by high resistance or an open condenser circuit. "Black" points are due to burning oil and grease, probably caused by excessive lubrication of the cam.
- *Sparkplugs:* These should be cleaned, tested and the gaps set according to "specs." Replace plugs which have cracked insulators or loose electrodes.
- *Carburetor:* It must be removed and thoroughly cleaned in solvent. Air-dry before reinstalling. Adjust choke and throttle linkages. Make final adjustments on carburetor after inspecting electrical system and other fuel system supply components.

• *Generator and starter:* Remove covers and examine brushes for proper bearing and pressure on commutator. Check brush spring tension. Brushes should have at least 60% bearing surface. Use only factory recommended brushes, and see that "pigtails" do not restrict free action of brushes and do not short against frame or cover. Commutator should be bright, clean and free from pitting. Slots between bars should be clean, and the mica separations kept below the level of the commutator bars.

• *Lift-and-tilt mechanism:* Inspect and equalize lift chains. If chains are defective, replace the entire chain. Do not replace sections. Check lift bracket rollers and sprockets for smooth, free operation. Check glands, packing boots and mounting pins on tilt cylinders. Check packing boots and vents on lift cylinders.

• *Steering and wheel alignment:* Make certain tires are properly inflated before checking steering or wheel alignment. Steering connections should be checked, adjusted and lubricated before wheels are aligned (truck weight should be on the wheels).

• *Contactors (electric-powered truck):* With the cover removed, examine contactors for proper functioning. Place controller lever in first speed, depress pedal and note whether contacts are breaking and closing properly. Examine the contactor tips for burning and wear. Some transfer of metal is normal, and if it does not interfere with the closing of the circuit, let it alone. Use a spring scale to correctly adjust spring tension according to maker's recommendations.

• *Controller:* Remove cover and see that controller fingers are making proper contact with the drums. All fingertips should be in exact line and evenly spaced on neutral center. Inspect contact tips and segments for pitting. If tips are badly pitted, it is probably due to weak finger tension. Check tension by lifting each finger off controller and seeing if it snaps back freely and quickly.

Six-Month Inspection

The 1,000-hour or semi-annual inspection includes examination of the slower wearing parts and other items not routinely inspected weekly and monthly.

• *Wheel bearings:* Remove wheels and examine bearings for cracks, broken rollers and excessive wear. Replace as required and repack with recommended lubricant. Also clean and repack spur gear and ring gear on drive wheels.

• *Brakes:* Inspect brake linings for dirt, grease and wear. If they are excessively dirty, wash in carbon tetrachloride or other solvent. Dry before replacing. If linings are glazed, score them with a wire brush. If the lining is worn, it should be replaced before the brake drums are scored. Inspect drums to make sure they are not scored or rough. Where brakes are timed with an electrical interlock to main travel controller and contactors, care must be taken not to disturb the timing when adjusting the brakes.

• *Lubrication:* Drain transmission and differential gear boxes when hot. Be sure to clean magnetic drain plugs of all metallic particles before reinstallation. Fill with recommended lubricant (usually SAE 90).

• *Hydraulic system:* Drain system by disconnecting high-pressure line from hydraulic valve and operate engine until entire system is drained. Reconnect the high-pressure line and refill the system with manufacturer-recommended hydraulic oil. In any case, oil used should have a low pour point and contain rust inhibiting and non-foaming characteristics. Check valve pressure and pump pressure with gage.

• *Chains and forks:* Check forks for even mounting. A 1/4-inch difference at the ends is permissible.

Finally remove chains and clean in kerosene. Dip in engine oil (SAE 10). Drain, then install and adjust.

• *Preventive maintenance programs pay:* Numerous studies have shown that programs incorporating the inspections outlined above, have reduced overall fork-truck maintenance costs from 15% to 25%.

III. ATTACHMENTS

There are a great variety of attachments which can be installed on or used with a lift truck to perform particular tasks. A few of the more popular attachments are discussed below.

In selecting the basic lift truck, it should be noted that the use of attachments will lower the rated capacity of many trucks and sometimes increase aisle width requirements for intersections and for right-angle stacking.

A. Reach Mechanism

In 1963, the author designed a special drive-in rack which, when used with a reach attachment to a conventional fork-lift truck, increased storage capacity by 40%. This innovation, the first such application of the drive-in rack concept, was covered in detail in the September 1963 issue of *Transportation and Distribution Management* magazine. An investment of $10,000 in drive-in racks plus $8,000 for a reach-truck enabled the company to defray indefinitely a capital expenditure of $250,000 for warehouse expansion. By the time (after five years) that the company's growth required more warehouse space, other logistical considerations precluded expansion at the existing site.

Reach mechanisms are used primarily with straddle-type trucks to avoid entrance of outriggers into floor storage positions when placing or retrieving loads. They extend up to 52 inches and are also used to reduce aisle width requirements in rack storage areas. The nominal load capacity limit of the reach attachment is 4,000 pounds.

B. Clamps for Cartons and Rolls

A major, nationally organized retail chain applied clamp attachments to handling major appliances. Special cartons were designed which then eliminated the need for pallets and minimized storage space requirements. The use of the clamp attachments saved the company $150,000 per year.

In using clamps, the cartons must form uniform patterns with adequate interlocking. Clamp attachments are unsuitable for mixed load carton sizes.

Roll clamps are used throughout the paper industry and for bale handling. A major publishing company first tried roll clamps in one of its smaller divisions with a resulting $25,000 annual savings in materials handling labor. The program was subsequently expanded to the company's other divisions.

C. Rams

Ram attachments are best used with items having center cores such as reels of wire, linoleum rolls, carpeting and so on. Be careful in selecting ram diameter since tight fits slow down productivity.

D. Sideshifters

These attachments are used primarily for putting stock into or retrieving stock from narrow-aisle rack configurations. They are also used to align loads in trailers and railroad cars.

E. Load Rotators

This category includes a variety of devices on the market which are designed to tilt, turn, dump or invert special handling units.

A sanitary products distributor, for example, uses a rotator attachment to retrieve 55-gallon drums from bulk storage, then lift and dump the contents into large vats.

F. Turnafork Attachments (90-Degree Rotating Forks)

These are used for tier-picking operations and as an interface between pallet and palletless handling. They also serve as regular forks and as tier-clamping devices when equipped with sideshifters.

G. Platforms

Special permanently installed or removable platforms are available for off-the-floor order selection. Lift trucks may be equipped with dual controls for both standard driver position operation and operation by a man on the platform.

Removable platform units are useful for intermittent operation—i.e., day-shift stock placing (without platform) and night-shift order selecting (with platform)—as an inventory taking stand, or for maintenance work.

H. Push-Pull Mechanisms

These devices are gaining in popularity for palletless operations with "slip sheets." They are primarily used in trailer or rail-car loading and unloading operations.

IV. SELECTION OF STANDARD PALLET RACKS

A significant amount of money can be saved by a knowledgeable rack purchaser. The purchaser (user) knows the application best, and by going through the analysis and detailing of rack requirements he is able to achieve the best utilization of existing space. Also, this intimate involvement helps in avoiding costly mistakes.

CASE IN POINT: A cosmetics manufacturer, desiring to build a new and more efficient 150,000 square foot distribution warehouse, asked several consulting firms to bid on developing a layout and preparing specifications covering materials handling equipment.

The low bid submitted was $85,000. Management decided to permit two of its own Manufacturing department industrial engineers, who had no prior warehouse experience, to do the job. Using the information listed below and equipment selection guidelines listed elsewhere in this book, the work was completed in six months at a cost of $10,000.

The material presented in the following pages covers the selection and preparation of specification information data for standard pallet racks. It includes a checklist of preparation data required to define rack requirements, layout consideration and a standard functional specification format. The information is applicable to any company and can be applied by anyone willing to make the effort. No prior materials handling background is necessary.

A. Checklist of Preparation Data

1. *Pallet information*
   a. Length, width and nominal height.
   b. Type construction.
2. *Load dimensions*
   a. Maximum length (or depth) including pallet overhang.
   b. Maximum width including pallet overhang.
   c. Maximum height including pallet.
3. *Maximum load weight per pallet*

   This includes consideration of "wet weight" if product is subject to water soaking by sprinkler systems.
4. *Rack data*

   Pallet racks are usually specified by "section." The following "section" information must be covered in specifications.

   a. *Number of openings (pallets per bed or tier)*

      Standard pallet racks are usually ordered as double openings; i.e., two pallets per bed. For pallet weights in the 2,000- to 5,000-pound range, double opening sections present reasonable limits for structural steel members. Light pallet loads offer an opportunity to increase the number of openings.

   b. *Weight per bed or tier*

      Multiply the number of openings by maximum load weight per pallet (including pallet weight).

c. *Rack section depth*

Single pallet deep storage is the most popular rack configuration. However, the development of extended reach-fork-truck equipment permits the use of a "double depth" rack. In this instance the rearmost pallet is not accessible without removing the "aisle" pallet.

d. *Rack section width (distance between uprights)*

Adequate clearance for placement and retrieval of pallets must be allowed in the width specification. Four inches is recommended between loads and between loads and uprights.

Special consideration must be provided for two common handling problems:

(1) High-elevation storage. For bed heights above 15 feet, 6 inches should be allowed to ease driver alignment problems.

(2) Straddle truck use (no reach attachment). Straddle trucks must enter the floor space (first tier) with their outriggers. Since outriggers are usually a minimum of 4 inches wide and the load is carried between the outriggers, a careful check must be made of clearances between uprights.

Note: Rack width does not include an allowance for upright member thickness. For multi-section width calculations in layout work, add upright dimensions.

e. *Rack height*

Sections are specified according to the number of tiers. Dimensions must be provided to cover total height as well as height between tiers or beds.

(1) Tier height is calculated by totaling three components:

(a) Load height including pallet.

(b) Lift clearance. A minimum of 4 inches is required to provide space to pick the pallet from the bed or place a pallet on the bed. Six inches clearance is preferred to speed up fork drive handling. Nine inches should be provided for tiers above 20 feet.

(c) Bed steel height. Total tier height (from top of one tier bed to the top of the next tier bed) must account for the bed beam. This is normally a 3- to 4-inch dimension depending on bed loads. For bed loads over 8,000 pounds, this height allowance should be calculated by the rack supplier.

(2) Total rack height is the sum of the tier heights covering the distance from the floor to the top of the highest bed. This dimension should be checked against existing or designed building clear storage height and should include the maximum additional height of pallets to be placed on the top bed plus a suitable lift clearance.

Rack heights resulting in 16 to 20 feet or more of storage will require the use of intermediate sprinkler levels. Review and establish added height requirements and rack configuration for intermediates with the FIA, rack manufacturer and your insurance carrier.

f. *Adjustable height*

Although rack tier or bed heights are nominally specified to allow for maximum load dimensions, all specifications should provide for adjustment of tier (bed) heights on 3-inch centers. This allows the rack to be adjusted to actual product or load needs to increase unit capacity and provide future flexibility.

g. *Rack bracing*

Combining individual rack sections in long aisles or in back-to-back arrangements, requires cross bracing to provide rack stability. Specifications should cover this situation by stating "sway brace as required" in both specification text and associated drawings. The supplier is then responsible for specifying adequate bracing.

B. *General Layout Considerations*

1. *Adequate aisle width*

Aisle requirements are determined by fork truck and load dimensions plus traffic density. Right-angle stacking requirements for a particular fork truck must be matched to rack requirements for aisle determination. In preparing layouts, show or designate aisles and racks based on the larger of either the rack or load dimensions; i.e., indicate rack aisles by showing load positions on the layout.

2. *Wall clearances*

Most warehouse designs require a minimum of 2- to 3-foot clear area around perimeter walls (due to fire and/or health regulations). Care should be exercised in specification of odd or short sections to meet this requirement.

3. *Column clearances*

A prime input to rack design is the consideration of rack placement and column spacing.

If rack areas are a minor part of the operation and/or column spacing is determined by other factors, care must be exercised to design racks so that unwarranted space losses do not occur. If, however, rack areas form a major part of the operation, and can be the determinant of column spacing, rack and column size design must be carefully considered and specified to provide maximum space utilization.

C. Specification Format

Pallet rack specifications should be presented in the following format. If different rack section configurations are to be included in the layout, separate specifications should be prepared for each configuration.

1. *Description:*

   a. *Section configuration*

      Number of openings, and number of tiers ("X" pallets per section, "X" wide by "X" high).

   b. *Pallet dimensions*

      Length and width of pallet in inches.

   c. *Adjustable height*

      Beds adjustable on "X" inch centers.

   d. *Section dimensions*

      (1) Depth of bed in inches from outer edges of bed.

      (2) Width of bed in inches between uprights.

      (3) Tier height = height in inches between tiers.

   e. *Bracing*

      To be specified by supplier.

2. *Load specifications*

   a. Maximum weight per intermediate bed in pounds.

   b. Maximum weight per top bed in pounds.

   c. Total section load in pounds. For ease of fork truck handling it is recommended that the bottom pallets be on beams—off the floor. If floor pallets are used, their weight should not be included.

3. *Number required*

   Number of sections.

D. Sample Specification of a Standard Pallet Rack

1. *Description*

   a. Double opening, four-tier (eight pallets, two wide by four high, including floor pallets).

   b. Pallet size = 48 inches by 40 inches; four-way, flush stringer pallet.

   c. Adjustable bed heights on 3-inch centers.

   d. Section dimensions:

      (1) Depth of bed = 36 inches.

(2) Width of bed = 94 inches.
(3) Tier height = 59 inches (includes bed beams).

e. Sway brace as specified by supplier.

2. *Load specifications*

a. Maximum weight for intermediate beds = 4,000 pounds.
b. Maximum weight for top bed = 8,000 pounds.
c. Total section load = 16,000 pounds (excludes floor pallets).

3. *Number required*

175 sections.

V. WHEN TO TRADE A FORK-LIFT TRUCK

Too frequently fork trucks are traded or sold only when they have become either obsolete or so dilapidated that the required action is painfully obvious. This situation is in marked contrast to the way in which most firms determine the trade-in point for delivery trucks, overland cabs and executive cars. In the latter cases, detailed economic studies are usually conducted and decision rules established.

Don Adams, Director of Materials Handling and Warehousing for U.S. Plywood Champion Papers, Inc., has developed the best system yet seen to determine when to trade in fork trucks and achieve the lowest lifetime cost. The system requires detail cost and other record keeping and reporting at the fork-truck user level. Once reporting discipline is achieved, the computer takes over and the rest is easy. An important feature of the system, is, however, that it lends itself to manual as well as computer analysis. It is, therefore, applicable to the small as well as the large company.

Don's system is designed to cover the company's 7,000 materials handling vehicles at more than 200 locations. A separate department was established to collect and maintain cost records for these vehicles. Each quarter, the computer prints out (see Exhibit 9-1) a cost history report for each unit. This report shows maintenance, depreciation and total cost per hour. The "lowest" total cost point, can be readily seen. That is the time to trade or sell the unit. To make evaluation even simpler, the computer also generates a total cost curve each quarter (see Exhibit 9-2). After six quarters of data have been accumulated, the computer projects an estimated "lowest total cost point."

Both exhibits refer to the same vehicle. In Exhibit 9-1, the last column shows the total cost per hour to operate the fork truck. Notice the significant increase in this column to $1.11 per hour in the first quarter of 19y3. The exhibit shows that the cause of this increased cost was in maintenance expense. In Exhibit 9-2, the computer has projected a total cost schedule out to 19y5. Note that the recommendation is to trade the vehicle in the first quarter of 19y3.

The reporting system just described also provides important operating information; i.e., percent of equipment utilization and quarterly maintenance costs. If, for example, a division requests additional equipment, the first item to check is the per-

```
**UNIT # 0226 ** CODE 100101C 3512     CHAMPION INTERNATIONAL CORPORATION          MATERIALS HANDLING                 PERIOD ENDING 03-31
                                          REPORT ON EQUIPMENT USAGE       DEPT: MOBILE EQUIPMENT                 PAGE: 20 DATE: 05/18

GROUP PAPER & ALLIED     DIVISION PAPER            SUB-DIV MANUFACTURING                          LOCATION HAMILTON           DEPT: 0855

CLASS A-HBB COMB ENG FORK TRUCKS CUSHION TIRE VALVED LPG         MFG HYSTER           TRANS AUTOMATIC        CAPACITY  5000     SHIFT C

MODEL NUM S50B        SERIAL NUM C2D-5593M         PRICE     9,345 OWNSHIP COST  7.0   RECEIVED 11/15/x8 STARTS 4-x8                DAYS 7

      FEATURES, SPECIAL EQUIPMENT & COMMENTS -    1 -MAST  70/150/ 39 FORKS                       HM X OHG X LBR   * PS X ACCUM
                                                  2 -
                                                  3 -                                                                   HM 064
                                                  4 -PO#40743  TRADE IN  -0079  -      VALUE $2600                 ARA#9534
```

QTR	QUARTERLY HOURS USED	QUARTERLY PERC UTIL	QUARTERLY MAINT COST	QUARTERLY COST/HOUR	CUMULATIVE HOURS USED	CUMULATIVE MAINT COST	CUMULATIVE DEPREC COST	CUMULATIVE TOTAL COST	COST PER HOUR MAINT	COST PER HOUR DEPREC	COST PER HOUR TOTAL	COMMENTS
4-x8	417	19.1	26	.062	417	26	654	680	.062	1.57	1.63	
1-x9	1,147	52.5	249	.217	1,564	275	1,262	1,537	.176	.81	.98	
2-x9	756	34.6	166	.220	2,320	441	1,828	2,269	.190	.79	.98	$16.14 MAL TREAT
3-x9	666	30.5	198	.297	2,986	639	2,354	2,993	.214	.79	1.00	X
4-x9	728	33.3	184	.253	3,714	823	2,843	3,666	.222	.77	.99	
1-y0	837	38.3	391	.467	4,551	1,214	3,298	4,512	.267	.72	.99	
2-y0	809	37.0	349	.431	5,360	1,563	3,721	5,284	.292	.69	.99	$18
3-y0	741	33.9	429	.579	6,101	1,992	4,115	6,107	.327	.67	1.00	$329 DAMAGE TO HOIST CYL SHELL
4-y0	639	29.3	221	.346	6,740	2,213	4,481	6,694	.328	.66	.99	$391 DAMAGE TO HOIST
1-y1	694	31.8	239	.344	7,434	2,452	4,821	7,273	.330	.65	.98	$24
2-y1	684	31.3	425	.621	8,118	2,877	5,138	8,015	.354	.63	.99	$310 MALTREAT TO UPRIGHT
3-y1	547	25.0	339	.620	8,665	3,216	5,432	8,648	.371	.63	1.00	
4-y1	619	28.3	412	.666	9,284	3,628	5,706	9,334	.391	.61	1.01	$9 DAMAGE
1-y2	650	29.8	546	.840	9,934	4,174	5,961	10,135	.420	.60	1.02	
2-y2	519	23.8	314	.605	10,453	4,488	6,198	10,686	.429	.59	1.02	$284 DAMAGE TO HOIST CYL
3-y2	500	22.9	272	.544	10,953	4,760	6,418	11,178	.435	.59	1.02	$6
4-y2	332	15.2	335	1.009	11,285	5,095	6,623	11,718	.451	.59	1.04	REPAIRS TO ENGINE & ELEC SYS
1-y3	450	20.6	1,173	2.607	11,735	6,268	6,814	13,082	.534	.58	1.11	$338 NEW CAM LEVER CONV PLATES

RESIDUAL VALUE OF UNIT 2,531

**UNIT # 0226 **

Exhibit 9-1

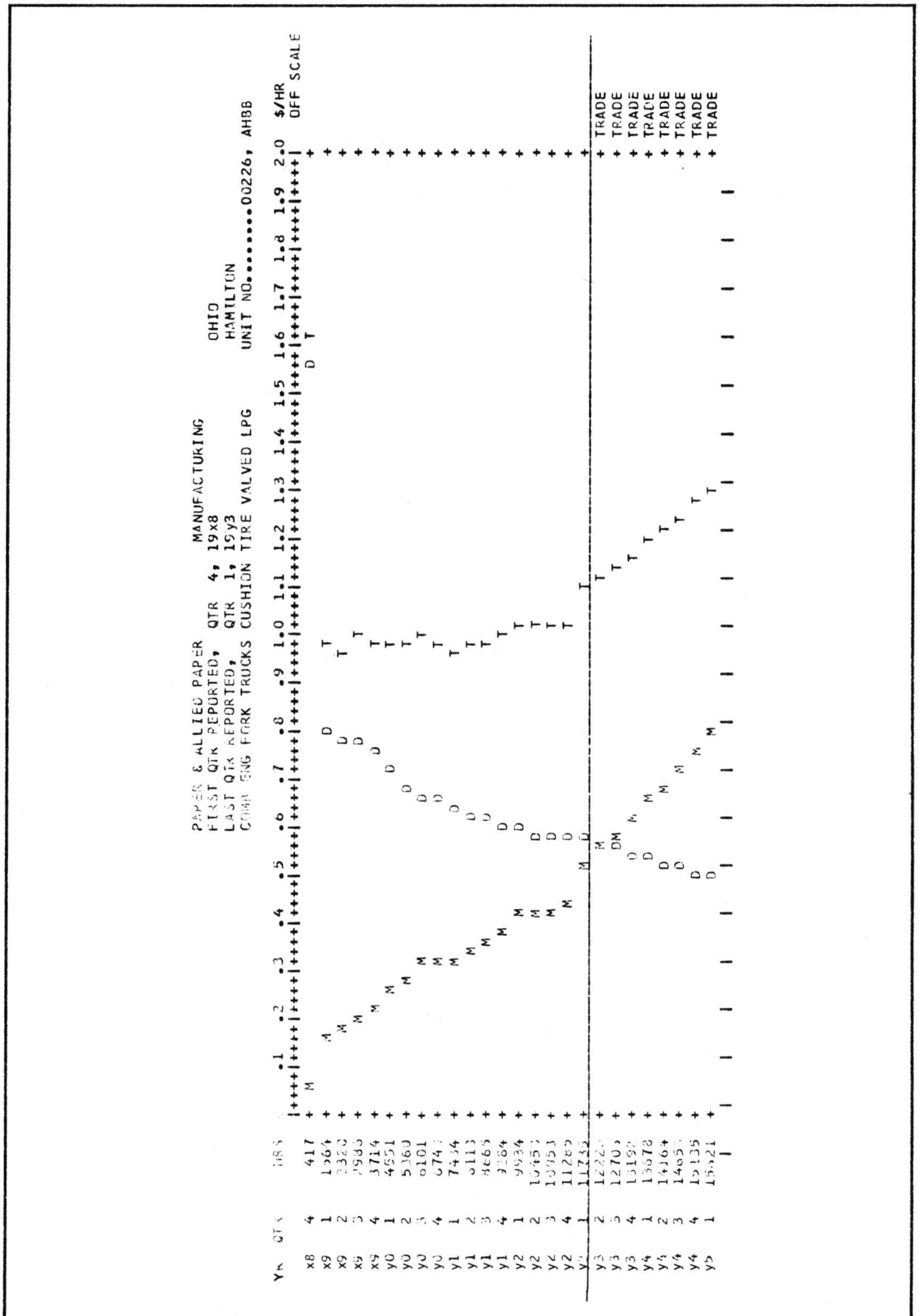
PAPER & ALLIED PAPER
MANUFACTURING
FIRST QTR REPORTED, QTR 4, 19x8
LAST QTR REPORTED, QTR 1, 19y3
COMB ENG FORK TRUCKS CUSHION TIRE VALVED LPG
OHIO
HAMILTON
UNIT NO.........00226, AHBB
$/HR
OFF SCALE
TRADE

Exhibit 9-2

cent utilization for similar units at that location. If utilization has been running at a high rate, additional capital equipment expenditures may be justified.

System Savings: Champion's studies have shown that net savings related to the program, taking into account the increased capital investment as a result of lower fleet age and reduced maintenance costs, is 10% to 15% of total fleet costs. Other savings such as warranty adjustments, fleet size reduction, etc., amount to 2% of total fleet costs.

10

Surveying, Fact Gathering, Analyzing, Recommending and Implementing Inventory Management Systems

Only a relatively few companies have been able to realize the full profit potential from their inventory management systems. One of the major problems has been poorly prepared system implementation programs. This chapter outlines the basic requirements needed to successfully implement inventory management systems.

THE INVENTORY SURVEY—A SYSTEMATIC APPROACH

It is strongly recommended that a detailed survey of a company's inventory management position be conducted before changes are made to existing procedures.

The survey serves several important purposes.

- It provides an opportunity for all departments to express their needs, and how well the existing procedures satisfy those needs.
- Top management uses the survey as a means to review policies which affect inventories; i.e., service level objectives, credit approval procedures, handling of obsolete merchandise, handling of rush orders, definition of what constitutes a back-order, desired order-cyle time and many others.
- It focuses corporatewide attention to the inventory management area.
- The company learns, often for the first time, exactly where it stands in regard to several important performance areas:

(a) Customer service level.
(b) Inventory investment trends.
(c) Inventory turnover and weeks of supply.

During the survey, flow charts are prepared covering key departments and procedures. Often this analysis results in the development of more efficient methods of operation.

- Simulation of new approaches to inventory management can be conducted to determine which, if any, system is most appropriate to the firm.
- Proposed system costs are determined and compared with current costs. A better system may be desirable solely on the basis of improved customer service, but economic justification is important.
- Customers are often interviewed during the survey. Their perception of the company's service performance can be very revealing. One firm found, for example, that any reduction in order cycle time would be welcomed by customers but would not result in increased purchases. The survey of the company's internal procedures showed that the order cycle time could be reduced by three days and that using this time to consolidate shipments at the warehouse would save $125,000 per year in freight costs (all shipments were prepaid). Based upon the intelligence on customer attitudes obtained during the survey, management chose to consolidate its shipments and enjoy the resulting savings.
- The survey should conclude with a detailed plan for implementing any proposed changes. This clarifies the roles of all the participants, thus giving greater assurance of success.

A typical inventory management survey would follow the outline below:

1. Survey objectives.
2. Summary of findings.
3. Analysis of current inventory position and procedures.
4. New approaches to managing finished goods inventories.
5. What must be done to make the system work?
6. An action program.

Listed below are the highlights which would typically be shown under each of the sections of the survey shown above:

Survey objectives

(a) Evaluate inventory procedures to determine adequacy in terms of customer service needs and factory cost optimization.
(b) Define customer service level objectives.

(c) Evaluate effectiveness of the inventory management organizational structure.

Summary of findings

(a) Review of past three years shows no measurable progress made towards reducing inventories.

(b) Seasonal pattern of sales data makes moving average calculations necessary to observe trends in both sales and inventory data.

(c) Data indicates a product mix problem—overproducing on some items while underproducing on others.

Analysis of current inventory position and procedures

(a) Exhibit 10-1 shows a graph which presents a three-year view of a company's "Weeks of Inventory."

(b) Exhibit 10-2 presents a view of a "Total Finished Goods Inventory" over three years.

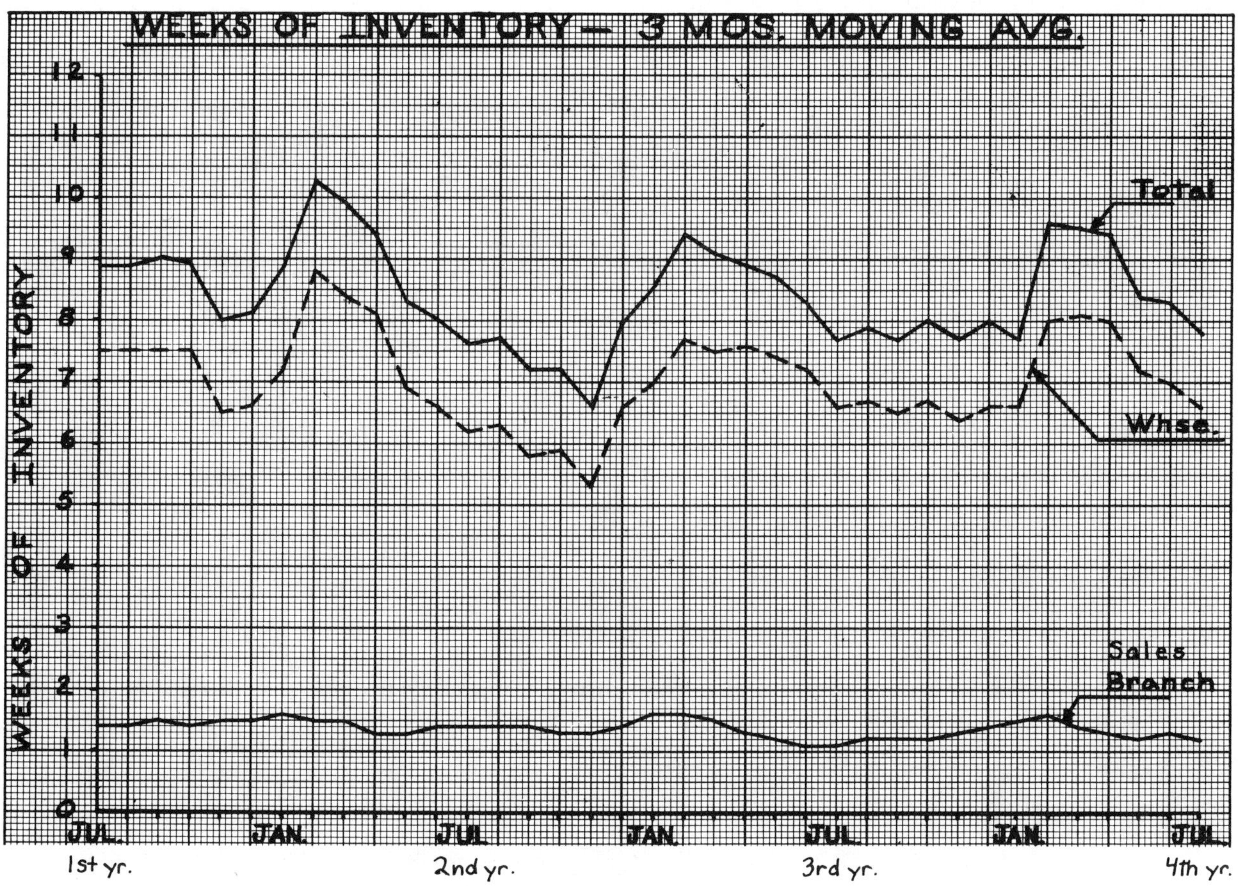

Exhibit 10-1

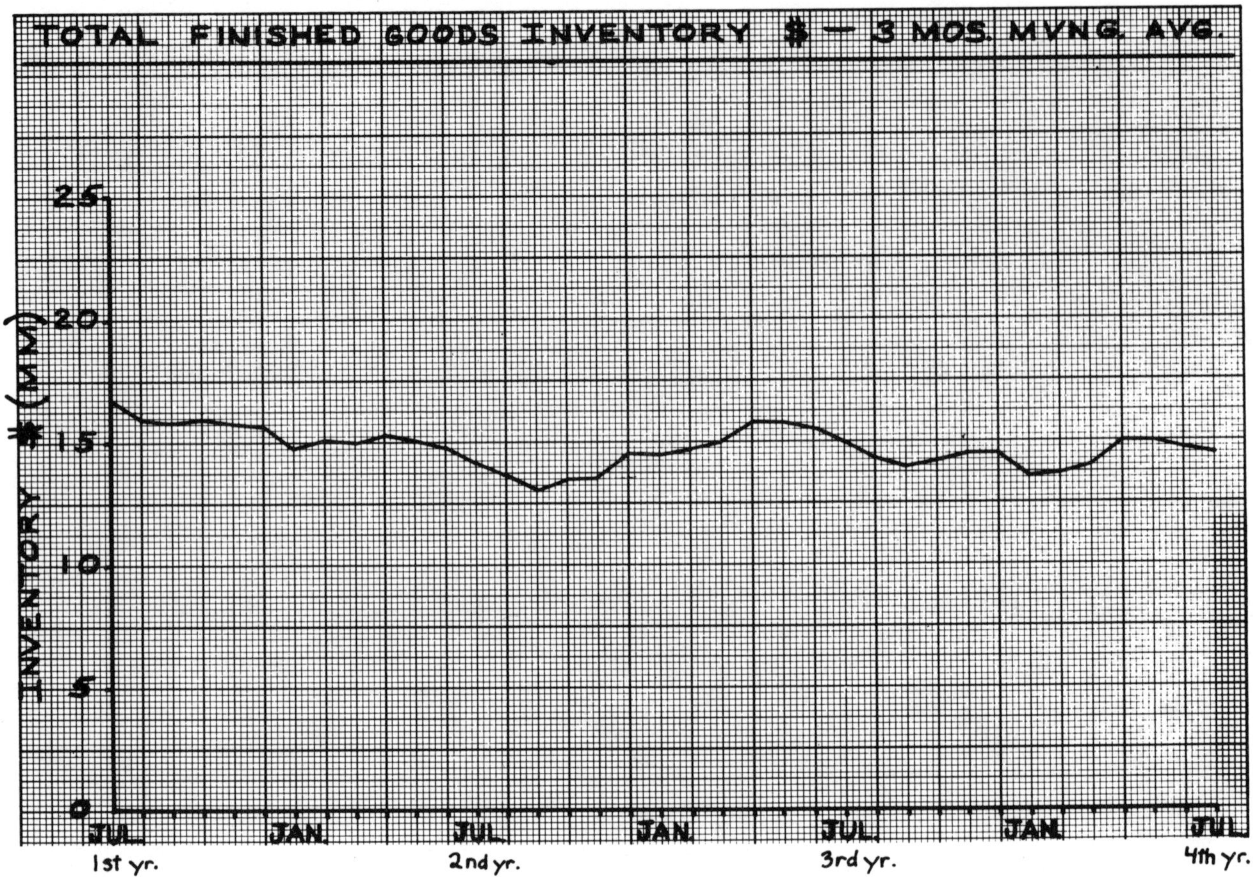

Exhibit 10-2

(c) Exhibit 10-3 shows the relationship between sales and inventory dollars. The closer the inventory cost curve is to the sales curve, the better inventories are being managed. This would demonstrate that inventory dollars are where they will bring the most return.

(d) Exhibit 10-4 presents those factors which should be included in a coverage of a company's "Existing Service Level Analysis."

(e) Exhibit 10-5 represents a typical procedural flow chart. Such charts indicate what inventory reports exist within the existing system and which departments generate and receive pertinent data. Dotted lines reflect informal communication links. Use of flow charts helps to prevent overlooking data requirements, formal and informal communication links in designing more automated systems.

(f) Exhibit 10-6 shows a typical A-B-C analysis based on "Sales and Inventory," and also on "Gross Profits." This example from a hardware wholesaler shows

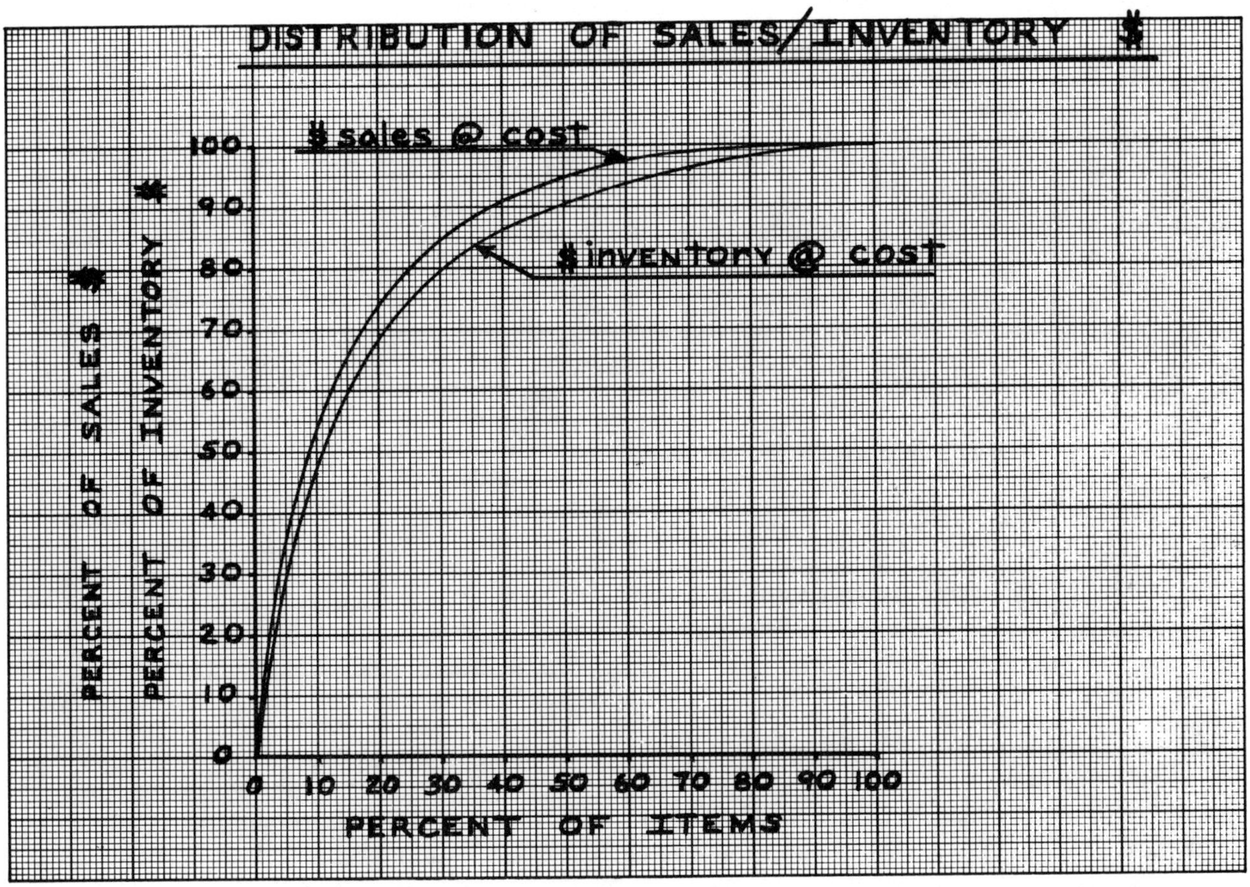

Exhibit 10-3

that a very small percentage of the items account for the great bulk of the gross profits. The analysis suggests that the company is carrying many obsolete items in its line. The report did, in fact, motivate management to close out and sell $1.75 million in obsolete inventory and then move on to modernize its product line.

NEW APPROACHES TO MANAGING FINISHED GOODS INVENTORIES

While finished goods usually receive most of the attention in the application of new inventory management systems, it should be remembered that manufacturing firms have other classes of inventory where significant inventory savings may be effected; i.e., raw materials, supplies and work in process. In fact, more money is usually tied up in the latter classes of inventory than in finished goods. Still, it is best to start with finished goods since the mix and absolute amount in that category determines what will be required in the other inventory classes.

EXISTING SERVICE LEVEL ANALYSIS		
% LINE ITEM SATISFIED ON TIME	=	96.3
% UNITS SATISFIED ON TIME	=	96.9
% ORDER FILLED ON TIME	=	75.6
% $ VALUE	=	98.5
AVERAGE LINES/ORDER	=	12.2
UNITS/ORDER	=	370.88
$ VALUE/ORDER	=	$2846

Exhibit 10-4

For most firms, a new approach means the introduction of a forecast system into the inventory management effort. This subject is covered in detail in Chapter 11. The survey should include a brief, non-mathematical discussion of the relation between forecasting and determining safety-stock levels.

WHAT MUST BE DONE TO MAKE THE INVENTORY MANAGEMENT SYSTEM WORK?

The discussion here should include both the *do's* and the *don'ts,* viz:

(a) *Try to avoid the following pitfalls:*

(1) Significant deviations from production and purchasing plans.

(2) Large forecast errors due to:

- Inaccurate or unknown promotional data.
- Poor estimates or sales on new products.
- Unplanned surges in orders from quarter ending salesmen's commissions and other pressures.

(3) Major, unplanned, short-term shifts between inventory locations; i.e., plant, warehouse, company-owned sales branch, salesmen's car stocks.

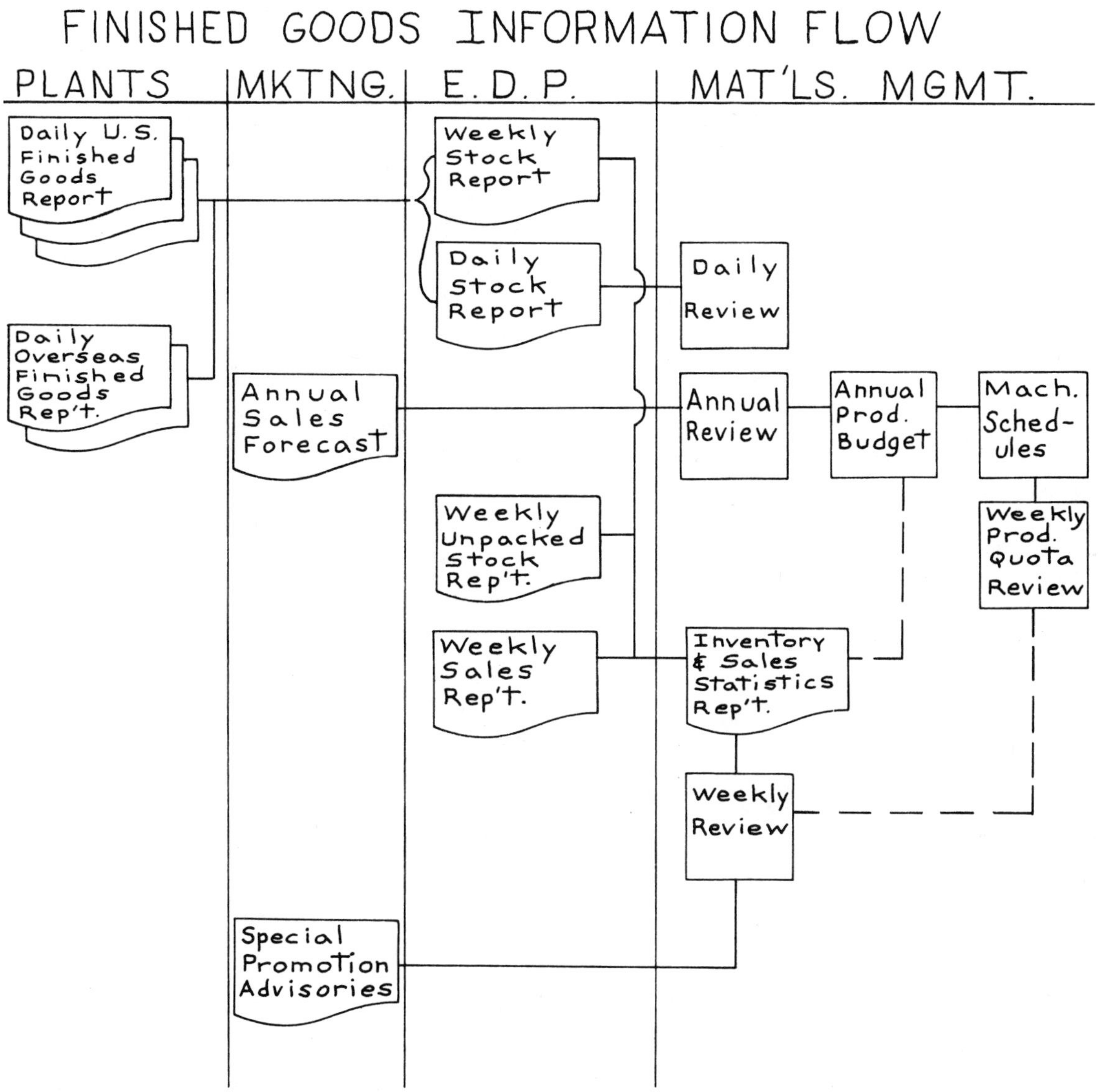

Exhibit 10-5

(b) *Try to do the following:*

(1) Thorough training and understanding of the revised inventory management system by all functionally related departments.

(2) Strong leadership and coordination, with each department fulfilling its responsibility.

(3) Provide advance information on promotions.

(4) Control over input data through timeliness of reporting with clear and uniform definition of terms.

SALES AND INVENTORY ANALYSIS

Category	% Items	# Items	@ Cost	% Sales	Inv. @ Cost	% Inv.	T/O	Mos. Supply
A	20	345	$6,476	92	$3,286	64	1.97	6.1
B	20	345	457	6	696	13	.66	18.2
C	60	1,034	82	2	1,167	23	.07	171
		1,724	$7,015		$5,149		1.36	8.8

GROSS PROFIT ANALYSIS

% Item	# Items	Gross Profit	Minimum G P per Item	% G P
10	714	$4,146 M	$672	92
10	738	220 M	168	5
80	5,596	114 M	0	3
	7,048	$4,480 M		

Courtesy: Schoolman Consultants Inc.

Exhibit 10-6

An Action Program

The "action" program should first include a diagrammatic representation of how the system will function (see Exhibit 10-7). It should then briefly state the long-run and the short-run programs which are planned to improve the firm's inventory management posture.

Finally, two implementation schedules should be shown. The first (see Exhibit 10-8) is a projection of manpower requirements by activity. The second should be a detailed PERT (Program Evaluation and Review Technique) chart showing all ac-

HOW DOES THE SYSTEM WORK

OPERATIONAL PROCEDURE

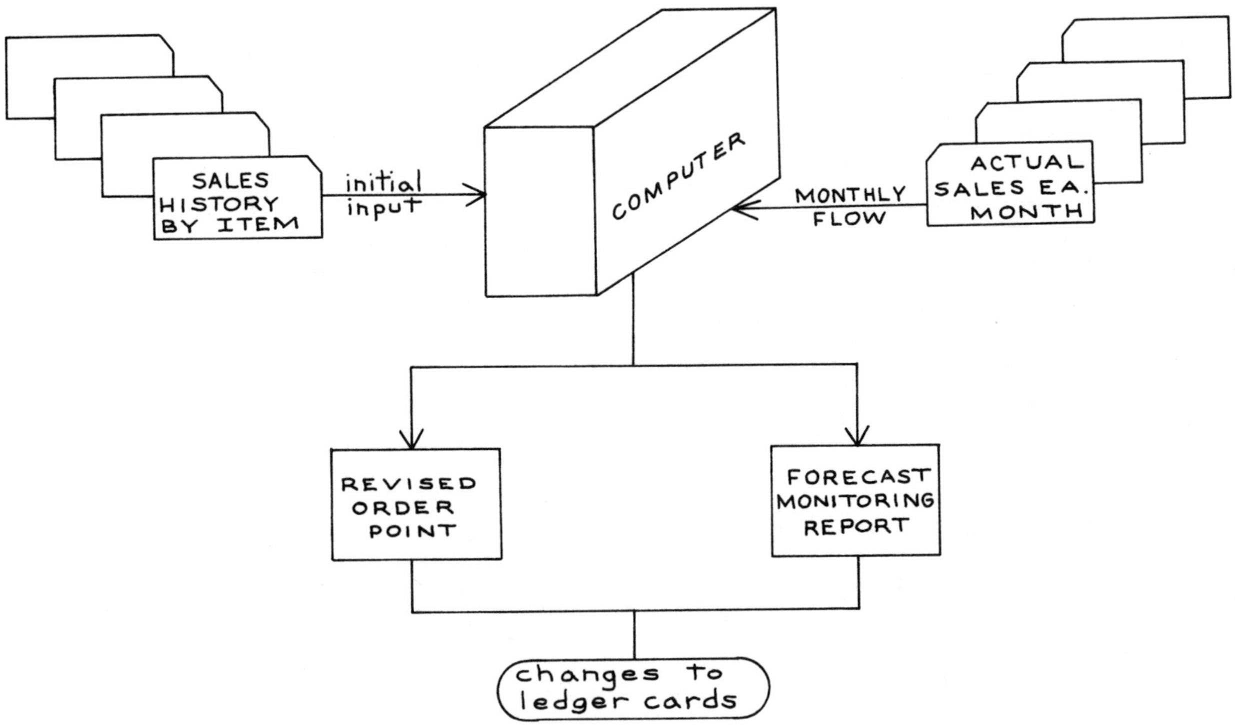

Exhibit 10-7

tivities which must be accomplished to achieve the implementation of the inventory management system (see Exhibit 10-9). Project scheduling helps in the following ways:

(a) Gives everyone working on project knowledge of where he fits in.
(b) Helps identify at beginning of project:
  a. All the work that has to be done.
  b. The priorities of the work.
(c) Helps to estimate the number of days elapsed and working time required to finish a job.
(d) On revisions, make more detail, eliminate.
(e) As project progresses, helps pull all things together.
(f) Makes management realize that implementation is a full-time responsibility.

INVENTORY MANAGEMENT SYSTEM

IMPLEMENTATION SCHEDULE

ACTIVITY	ELAPSED MAN WEEKS	# OF PEOPLE	MAN WEEKS
A. LAYOUT OF SYSTEM SPECIFICATIONS	2	2	4
B. INSTALL FORECAST SYSTEM	20	2	30
C. REDESIGN OUTPUT REPORTS	12	2	18
D. SYSTEM EVALUATIONS DEBUGGING	4	2	6
E. DOCUMENTATION (USER'S MANUAL)	2	2	3
F. TRAINING	3	2	6
	31 Wks.		67

INCREMENTAL SYSTEM COST - NEGLIGIBLE - $300 - mo.

TERMINAL FOR SIMULATION - $300 - mo. for first year.

TOTAL $600 - mo.

Exhibit 10-8

INFORMATION FEEDBACK SYSTEMS

Information feedback is the end result of the entire inventory management system. In implementing and then operating an inventory management system, you should be concerned with three levels of information feedback:

1. During the implementation phase, it is important to know if each part of the project (see Exhibit 10-9) as well as the overall effort is on schedule. A delay or lack of problem resolution in one area could have compounded delay and cost effects later. Throughout the implementation period, monthly progress meetings should be held and the PERT chart updated. Any changes in budget requirements will then be recognized early and "surprises" to top management avoided.

PERT CHART - FINISHED GOODS INVENTORY SURVEY

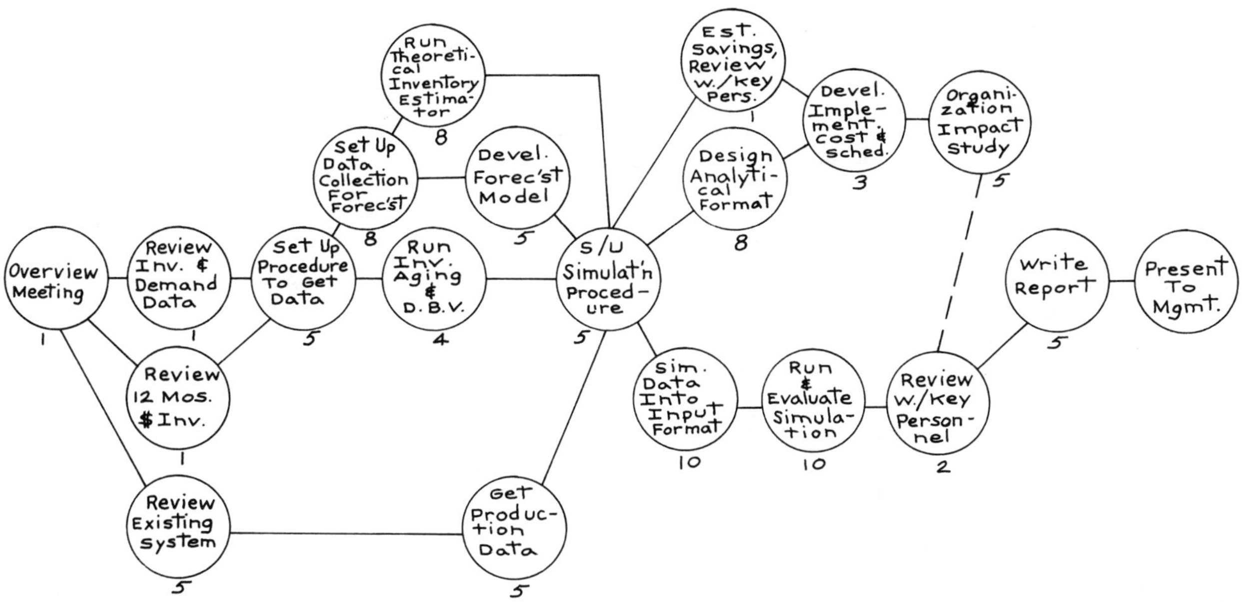

Estimated Elapsed Time = 93 Days

Exhibit 10-9

2. During actual operation of the system, information feedback required for decision making should be timely and on an exception basis. The exceptions reports provide a quick reference to those items requiring immediate attention. Total reports should be generated separately for more detailed review and analysis. A good example of an exception type report may be seen in Exhibit 10-10, "Order Point Exception Report." Using the report to balance inventories between four distribution centers enabled the company to improve customer service by five percentage points while total finished goods inventories were reduced by $1.5 million. The report shows an order point and a weeks of supply on hand figure based on a weekly forecast. Whenever total inventory for an item at any location falls below the order point, the total available inventory at all locations where the item is carried is displayed. This report is produced daily with a weekly summary report generated showing the status of all items. The asterisks in the far right-hand column indicate the location where the inventory has fallen below the order point. Note that total inventory is the sum of amounts in-transit to distribution centers and non-committed inventories. Non-committed inventory is defined as, "finished goods on hand

V-006 ORDER POINT EXCEPTION REPORT DATE 06/02/ 4201021-01

PLT	PRODUCT CODE	DESCRIPTION	IN-TRANS INV	+ NON-COMM INV	= TOTAL INV	BK-ORDER	ORDER POINT	WKS OF SUPPLY	WEEKLY FCST	*
12	8010-060	EP BQT BOX 50BOY	0	7,350	7,350	0	1,400	10.5	700	
45	8010-060	EP BQT BOX 50BOY	0	8,500	8,500	0	9,600	1.7	4,800	*
46	8010-060	EP BQT BOX 50BOY	0	20,400	20,400	0	3,000	13.6	1,500	
12	8010-824	EP BQT VAC 25BOY	0	9,925	9,925	0	5,600	3.5	2,800	
45	8010-824	EP BQT VAC 25BOY	0	11,000	11,000	0	38,600	.5	19,300	*
46	8010-824	EP BQT VAC 25BOY	0	23,650	23,650	0	4,600	10.2	2,300	
12	8010-825	EP BQT VAC 25GIRL	0	41,400	41,400	0	4,000	20.7	2,000	
45	8010-825	EP BQT VAC 25GIRL	0	7,325	7,325	0	35,600	.4	17,800	*
46	8010-825	EP BQT VAC 25GIRL	0	34,975	34,975	0	3,200	21.8	1,600	
05	8240-030	EP BLT BOX 50	0	500	500	0	0	999.9	0	
12	8240-030	EP BLT BOX 50	0	10,800	10,800	0	94,800	.2	47,400	*
45	8240-030	EP BLT BOX 50	970,000	3,094,850	4,064,850	0	907,000	8.9	453,500	
46	8240-030	EP BLT BOX 50	0	258,400	258,400	0	255,400	2.0	127,700	
12	8240-125	EP BLT BOX 50VEND IMP	0	2,500-	2,500-	0	900	2.7-	900	*
05	8240-300	EP BLT 5 PKS 50	0	37,500	37,500	0	4,000	18.7	2,000	
12	8240-300	EP BLT 5 PKS 50	0	0	0	0	0	999.9	0	
45	8240-300	EP BLT 5 PKS 50	1,392,500	2,272,800	3,665,300	0	1,486,500	6.1	594,600	
46	8240-300	EP BLT 5 PKS 50	0	109,600	109,600	0	250,000	.8	125,000	*
05	8230-300	EP V-P 5 PKS 50	0	2,500	2,500	0	0	999.9	0	
45	8230-300	EP V-P 5 PKS 50	400,000	666,700	1,066,700	0	452,550	4.9	215,500	
46	8230-300	EP V-P 5 PKS 50	0	67,700	67,700	0	126,800	1.0	63,400	*
05	8020-030	EP P-F BOX 50	0	1,000	1,000	0	0	999.9	0	
45	8020-030	EP P-F BOX 50	964,000	2,283,250	3,247,250	0	729,330	10.2	317,100	
46	8020-030	EP P-F BOX 50	0	251,050	251,050	0	542,200	.9	271,100	*
05	8020-300	EP P-F 5 PKS 50	0	2,000-	2,000-	0	600	6.6-	300	*
45	8020-300	EP P-F 5 PKS 50	66,000	1,745,450	1,811,450	0	717,150	5.3	341,500	
46	8020-300	EP P-F 5 PKS 50	0	46,800	46,800	0	276,200	.3	138,100	*
12	8730-030	EP ALT BOX 50	0	55,850	55,850	0	12,400	9.0	6,200	
45	8730-030	EP ALT BOX 50	0	543,350	543,350	0	69,000	15.7	34,500	
46	8730-030	EP ALT BOX 50	0	14,400	14,400	0	35,200	.8	17,600	*
12	8730-570	EP ALT 7 PKS 70	0	61,320	61,320	0	20,600	5.9	10,300	
45	8730-570	EP ALT 7 PKS 70	0	827,050	827,050	0	183,000	9.0	91,500	
46	8730-570	EP ALT 7 PKS 70	0	13,650	13,650	0	69,200	.3	34,600	*
45	8640-359	EP MON 5 PKS 5070	0	0	0	0	1,000	.0	1,000	*
45	8030-030	EP FAV BOX 50	274,000	746,400	1,020,400	0	347,130	6.1	165,300	
46	8030-030	EP FAV BOX 50	0	63,750	63,750	0	128,200	.9	64,100	*

Exhibit 10-10

less orders received but not yet shipped." This provides a far more current picture of the company's inventory position.

3. The last type of information feedback is concerned with detecting significant item demand variations from normal. One of the most important reports in this category is the "Demand Filter Report." This report prints out all items which have had a significant variation from the usual demand pattern. It is usually generated once each month and would highlight any items which evidenced either a very high or very low deviation. If the variation were on the high side, the purchasing agent would evaluate whether a new order quantity should be calculated or whether a factor such as a special promotion was creating a one-time surge in demand. If the variation were on the low side, this might suggest that the order quantity should be cut to avoid getting caught with too much stock.

   Another report called a "Tracking Signal Report" provides information feedback on items which are trending up or down. This type of data is most useful to purchasing and production personnel.

A DISCIPLINED APPROACH IN PROCESSING INPUT DATA

Often overlooked is the fact that the inventory management system is only as reliable and useful a management tool, as its weakest link. And that weak link is usually located at the lowest levels in the organization.

A $4,000-a-year receiving clerk can, if he is negligent, cause inventory management problems. Night-shift employees who withdraw raw materials or supplies from stock without submitting the necessary documentation cause inventory system failure and stockouts. Shipping clerks who, through inattention, permit merchandise to be pilfered from the dock and then have substitute merchandise brought from the warehouse without submitting proper record of the fact, subvert inventory management systems. The list can go on and on. The situations described above are more the rule today than the exception. That is a hard fact which must be faced squarely. Frustrating yes, impossible no.

Let's take a look at several positive steps which can be taken to improve the quality of information processing:

MANAGEMENT PRESENCE

Inventory managers should get out into the warehouse and meet the men who pull the orders, drive the delivery truck, operate the fork-lift trucks, etc. People are disposed to do a better job of record keeping if they know you care; if they can relate their function to how it matters in the end. Try it. It works.

USE CYCLICAL INVENTORY COUNTING SYSTEM

Too many companies take physical inventories only once or twice each year. In the interim, many errors go undetected. Overstocking in some areas and stockouts in others is common.

Cycle inventory taking procedures work on a sampling basis. A small percentage of the total inventory is taken daily, weekly or monthly. More important items are counted more frequently. Inventory status errors are caught faster.

CONTROLLED ERROR TESTS

Here, receiving personnel, shippers, delivery drivers, checkers and other warehousemen are advised that deliberate errors will be introduced into their work as a test of their ability to catch mistakes. This procedure should not be used to "catch" people in an error. It is good preventive medicine. For example, a carton is added to a delivery truck some time before it departs. Does the driver return it at the end of the run? Suppliers usually cooperate too. They will, as instructed, periodically short ship on an order. Does the receiving clerk catch the shortage or merely accept the bill of lading?

USE CHECK-DIGITS IN ELECTRONIC DATA PROCESSING

Through mathematical construction, an additional digit is positioned at the end of a product code number which tests the validity of the code number itself. If used in the E.D.P. department, keypunching or written errors can be caught before data enters the inventory management system. This procedure minimizes the possibility of receipts or shipments being charged against incorrect products.

FOLLOW THROUGH ON ERRORS

A very helpful device in keeping reporting error frequency under control is the maintenance of error logs. Daily or weekly reports are maintained on critical data entry items. Individuals and/or locations responsible for the errors are indicated in the logs, and when the error rate is felt to be excessive, this is communicated to the responsible parties.

> *CASE IN POINT:* One multi-national consumer goods manufacturer found that his inventory management reports were becoming unreliable due to errors in reporting quantities en route from plants to warehouses. Sometimes the errors were in the amount of product either being shipped or being received. On other occasions, wrong destinations were recorded. It finally became so serious, that the Financial department was unable to balance its books at the end of the month. It was decided that an error log would be maintained and that the heads of the Distribution, Electronic Data Processing, Production and Accounting departments would meet once each week to review its status. At the beginning of this process, roughly 57 errors were being reported weekly. After two and a half months, errors were down to three per week. The log was continued but the weekly meetings were no

longer necessary. During that two and a half months, the executives learned a lot about their operations. Systems were improved, communication links established and respect for other people's problems established.

FUNCTIONAL AREAS WHICH SHOULD FALL UNDER INVENTORY MANAGEMENT ADMINISTRATIVE CONTROL

As was discussed in Chapter 2, there are many different corporate approaches and attitudes towards materials management organization structure and towards where the inventory management function might fit into the overall picture. In the case of J.C. Penney, for example, the Director of Distribution felt that inventory management should be handled autonomously at the store level. The position could also be taken that a central authority could provide inventory management guidance and could constructively evaluate individual performance.

While there are obviously a great many ways to structure the total materials management activity, some specific conclusions can be established from which a significant number of firms could benefit.

- In manufacturing companies, inventory management should be closely related to the production control function. Inventory management should have consolidated within its folds, responsibility for all classes of inventory, including raw materials, supplies, work in process and finished goods. Ordering strategies should be established which will permit the attainment of customer service level objectives at minimum inventory cost.
- In purely distribution organizations, inventory management should be combined with purchasing. Functional responsibility would include establishing order points and order quantities and the development of ordering strategies for finished goods which will permit minimizing inventory investment while maintaining desired customer service levels.

TOP MANAGEMENT SUPPORT

Much has been said on the subject of top management support for important projects, but it can never receive too great an emphasis in the inventory management area. This is the case because implementing inventory management systems requires crossing over many departmental lines of authority. Without strong backing from top management, it becomes infinitely more difficult to obtain data and the level of cooperation and participation required.

The inventory management survey, as outlined at the beginning of this chapter, is the most important device which you can employ to get the support you need from top management. Going through the information gathering and analysis effort, obtaining management input on policy matters and establishing budget criteria, all combine to give top management a feeling of confidence in your ability to get the job done.

GLOSSARY OF INVENTORY MANAGEMENT TERMS

ABC Analysis—A method of ranking stockkeeping units and setting priorities.

Back-Order—A stockout satisfied at a future time period.

Carrying Cost—The out-of-pocket expenses plus lost profit opportunities paid for each dollar in inventory.

Cash Commitment Schedule—A procedure for forecasting future working capital requirements due to changes in inventory levels.

Customer Service—See Service Level.

Demand Filter—A statistical method of filtering out unusual peaks and valleys in sales patterns.

EOQ-Economic Order Quantity—The order quantity minimizing total purchase plus carrying cost.

Exponential Smoothing—A statistical method of forecasting.

Forecasting—A systematic method of estimating future demand.

Inventory Aging Analysis—A method of identifying surplus, obsolete and overstocked items.

Lead Time—The total time between realizing it is time to reorder and the receipt of shipment including entering it into stock status records.

Mad-Mean Absolute Deviation—A statistical measure of forecast and/or sales variability about an expected value.

Opportunity Cost—Lost profit due to lack of working capital.

Order Cycle Time—Time from receipt of customer order to delivery of merchandise to the customer.

Order Point—The level of inventory at which a reorder should be placed.

Order Point System—A set of inventory decision rules that determine when to order as the inventory goes below a specified level.

Order Quantity (OQ)—The amount of product to purchase after it has been decided that an order should be placed.

Order Up-to System—A set of inventory decision rules such that orders are placed on a fixed time schedule. OQ's are placed so that the inventory will be brought "up-to" a specified level.

Perpetual Inventory Record—A systematic bookkeeping procedure for measuring inventory balances on hand or available for future orders.

Purchase Budget—Dollar restrictions on total purchases within a fixed time period.

ROA—Return on Assets—The percent return on a dollar invested in a product.

Safety Factor—A statistical measure used to calculate safety stock.

Safety Stock—Extra inventory allowing for variations in demand and vendor lead time guaranteeing certain customer service.

Service Level—1) The percent dollar demand satisfied within a fixed time period.
2) The percent of line items satisfied within a fixed time period.
3) The percent of invoices satisfied within a fixed time period.
4) The probability of a stockout within a reorder cycle.

Time Supply—The length of time an inventory position will last given a sales pattern.

Time Supply System—A set of inventory decision rules in which order points and order quantities are multiples of estimated demand. (Safety stock requirements are approximate.)

Tracking Signal—A statistical method of measuring forecast bias.

Turnover—The number of sales dollars per dollar of inventory investment. (Sales are usually measured at cost.)

Usage by Value—See ABC Analysis.

11

Planning: The Critical Element in Inventory Management

INVENTORY MANAGEMENT VS. INVENTORY CONTROL

The distinction between management and control of inventories is critical to the successful design and implementation of inventory *management* systems. It is a distinction all too often not made. Inventory control is merely that part of a system which enables you to keep track of what you have on hand and on order. Inventory management involves control but goes further; it is a system which enables you to achieve a pre-determined level of customer service with a minimum amount of dollars invested in inventory. Simply stated, the function of an inventory management system is to know when, what and how much to buy in order to achieve specific customer service and inventory turnover objectives.

Inventory management and inventory control systems vary in their degree of sophistication. Listed below are several of the most successful approaches to achieving inventory control and inventory management, from the simplest form to the more complex. Some inventory management systems use two or more forms to cover different parts of a company's inventory. The main point to remember is that there is no universal solution to the inventory management problem. Package systems applied elsewhere cannot simply be plugged in at your company and be expected to perform successfully. A good deal of expert tailoring is usually needed.

PERIODIC PHYSICAL COUNTS

This control system is found most frequently in small firms having few items in stock. The warehouseman or manager knows from experience at what physical levels additional quantities should be ordered.

PERPETUAL, IN-AND-OUT RECORD ON CARDS

The perpetual inventory card control system usually contains an order point or minimum quantity. When this level is reached, a purchase requisition is considered. This system reduces dependence on visual observations. Minimums are set by some knowledgeable manager without benefit of statistical analysis.

PERPETUAL + ON-ORDER STATUS ON CARDS

This control system is a refinement of the basic perpetual in-out record. It permits the inventory controller to quickly review the status of items which have previously fallen below the minimum levels. No separate record need be consulted.

BIN-TYPE DECISION RULES

Decision to reorder is based upon a pre-determined physical level.

TIME SUPPLY INVENTORY MANAGEMENT SYSTEMS

Time supply systems set order points and order quantities equal to a certain time supply for each product. For example, you might set your reorder point at a one-month supply and set reorder quantities at a two-month supply. What constitutes a month's supply is established through a statistical review of historical sales.

Introducing a degree of sophistication to time supply systems may be achieved by breaking the product down into A, B and C categories. "A" items might cover the top 20% of sales, "B" items the next 30% of sales and "C" items the lowest 50% of sales. The "A" items would receive the most analysis and attention.

Time supply systems are almost universally applicable and are fairly simple to introduce. It is important to remember, however that:

Order Points must equal Demand during (review time + lead time) + Safety Stock.

Safety stock may be set arbitrarily but usually is not more than a one-month supply, and you may often have too much or too little to cover individual stockkeeping units. Nevertheless, time supply systems represent a good way of getting into inventory management.

Time supply systems do not apply where quantity discounts are available. Here, Economic Order Quantity (E.O.Q.) systems are used. More will be said of E.O.Q. systems later.

Using a time supply system, it is fairly simple to determine what inventory turnover will be. Follow the two steps shown below:

1. Calculate Your Theoretical Inventory.

   Theoretical Inventory = ½ Order Quantity + Safety Stock

   If the O.Q. = 2 months' supply, and safety stock = 1 months' supply. then

   T.I. = ½ × 2 mos. + 1 mo.

   T.I. = 1 mo. + 1 mo. = 2 mos.

2. Calculate Your Theoretical Turnover.

   $$\text{Turnover} = \frac{\text{12 mos. sales}}{\text{Theoretical Inventory}}$$

   $$\text{T.O.} = \frac{\text{12 mos. sales}}{\text{2 mos.}} = \text{6 turns per year}$$

MIN.-MAX. SYSTEMS

Min.-Max. systems are used for low dollar value items where large quantities are stocked. Here, when a certain minimum inventory level is reached (called the order point), you order up to the maximum level. The maximum level would be the order point + the order quantity.

The min.-max. system requires some estimates of demand to permit establishing a minimum quantity or order point which will see you through the lead time period (length of time from reordering until receipt of those goods), plus something for safety stock. See Exhibit 11-1.

ECONOMIC ORDER QUANTITIES

For example, if a company expects to use 6,000 units a year at a unit cost of $2.50, orders can be processed in varying amounts. If a four-month supply is bought at one time, only three orders a year need be processed. If we assume that it costs $15 to process each order and 20% of the average inventory to carry it, we find that ordering costs are only $45 a year. This figure sounds good until we note that if we buy 2,000 units at a time the carrying costs on the inventory are $500 a year. The total costs of ordering and carrying are thus $545 a year. Since it is obvious that carrying costs may be reduced by ordering in smaller lot sizes, we can then go to the other extreme. If we order twice a month, we process 24 orders a year; at $15 an order, the cost is $360 a year. On the other hand, only $62.50 is spent on carrying costs, making a total cost of $422.50. This figure is somewhat better than $545 a year; however, it is obvious that we can do better by ordering somewhere between a four-month supply and a two-week supply. If we assume a purchase quantity of a two-month supply, we have six orders, which give us ordering costs of $90. The carrying costs are then $250, and the total costs are $340. This figure is somewhat less than either of the previous cost totals.

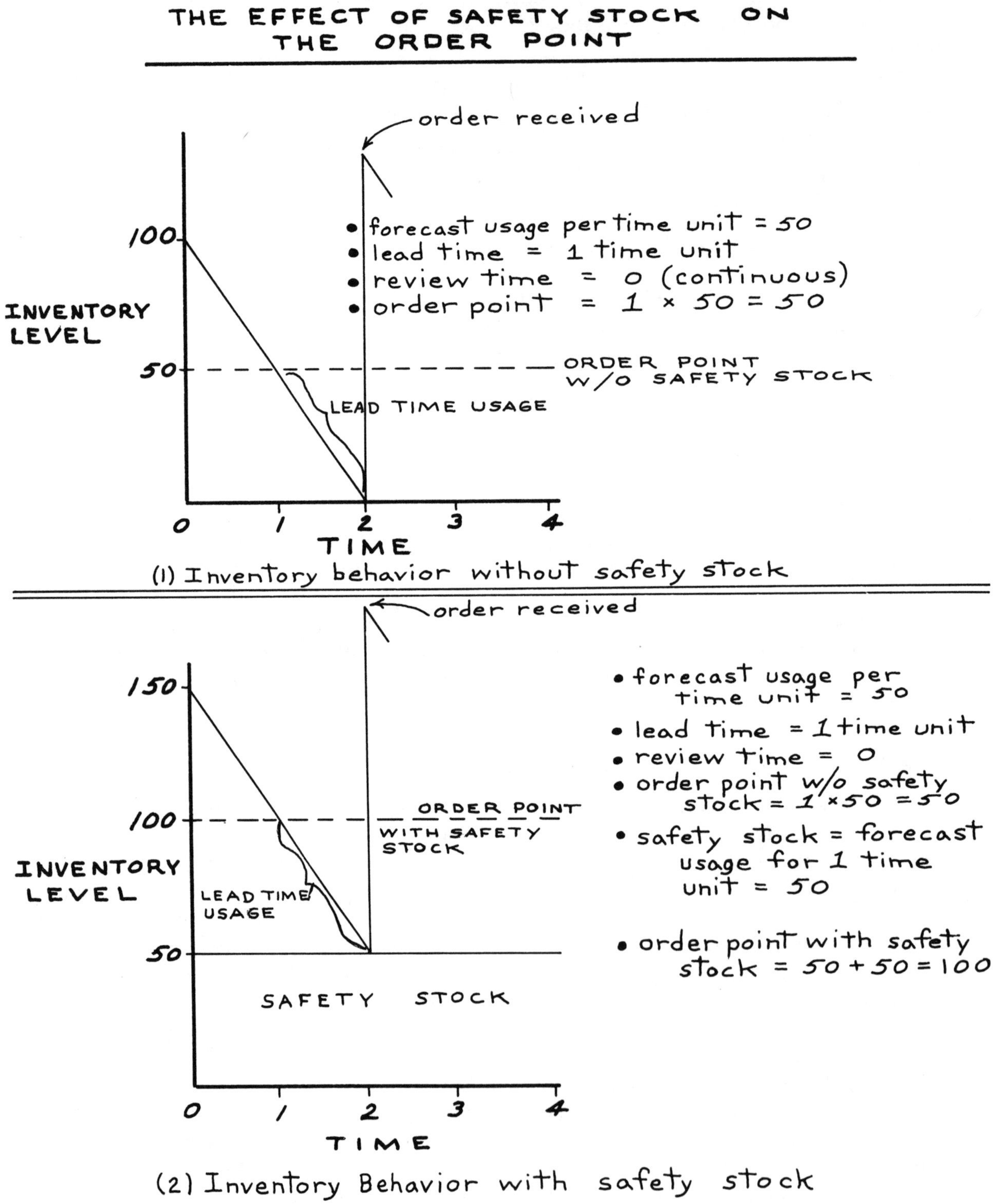
THE EFFECT OF SAFETY STOCK ON THE ORDER POINT
order received
100
50
INVENTORY LEVEL
• forecast usage per time unit = 50
• lead time = 1 time unit
• review time = 0 (continuous)
• order point = 1 × 50 = 50
ORDER POINT W/O SAFETY STOCK
LEAD TIME USAGE
0 1 2 3 4
TIME
(1) Inventory behavior without safety stock
order received
150
100
50
INVENTORY LEVEL
ORDER POINT WITH SAFETY STOCK
LEAD TIME USAGE
SAFETY STOCK
• forecast usage per time unit = 50
• lead time = 1 time unit
• review time = 0
• order point w/o safety stock = 1 × 50 = 50
• safety stock = forecast usage for 1 time unit = 50
• order point with safety stock = 50 + 50 = 100
0 1 2 3 4
TIME
(2) Inventory Behavior with safety stock

Exhibit 11-1

COST CATEGORY	3 ORDERS/YR. (2,000 UNITS)	24 ORDERS/YR. (250 UNITS)	6 ORDERS/YR. (1,000 UNITS)
A. Ordering Costs	$ 45.00	$360.00	$ 90.00
B. Carrying Costs	500.00	62.50	250.00
Total Cost	$545.00	$422.50	$340.00

It would be possible to continue trying new order quantities until the lowest total cost is determined, but such trial-and-error methods are time consuming and unnecessary. Hundreds of formulas have been developed for computing what is called the Economic Order Quantity; however, they all have essentially the same characteristics. By using a relatively simple formula:

$$Q = \sqrt{\frac{2P \text{ X } R}{I \text{ X } C}}$$

When

Q equals Economic Order Quantity
P equals Ordering Cost per Order
R equals Requirements per Year in Units
C equals Cost per Unit
I equals Carrying Cost in Percentage

The Economic Order Quantity for the illustration shown above can be determined, as follows:

$$Q = \sqrt{\frac{2 \times 15 \times 6000}{.20 \times 250}}$$

$$Q = \sqrt{3,600}$$

$$= 60 \text{ items at each order}$$

See also, Exhibit 11-2 for determination of E.O.Q.

If 600 items are ordered at one time, approximately 10 orders at a cost of $150 would be placed a year. The cost of carrying this inventory is computed as follows: The quantity ordered at one time (600) divided by 2 gives the average number of units on hand during the year. Multiplying 300 units by $2.50 gives the value of the average inventory ($750). Taking 20% of the average value of the working inventory gives the cost to carry that inventory ($150). The total cost of ordering and carrying this inventory is $300. No total cost figure less than $300 can be obtained under these cost estimates; therefore, 600 is the most economical order quantity.

Solving such a formula for every product every time the average monthly usage

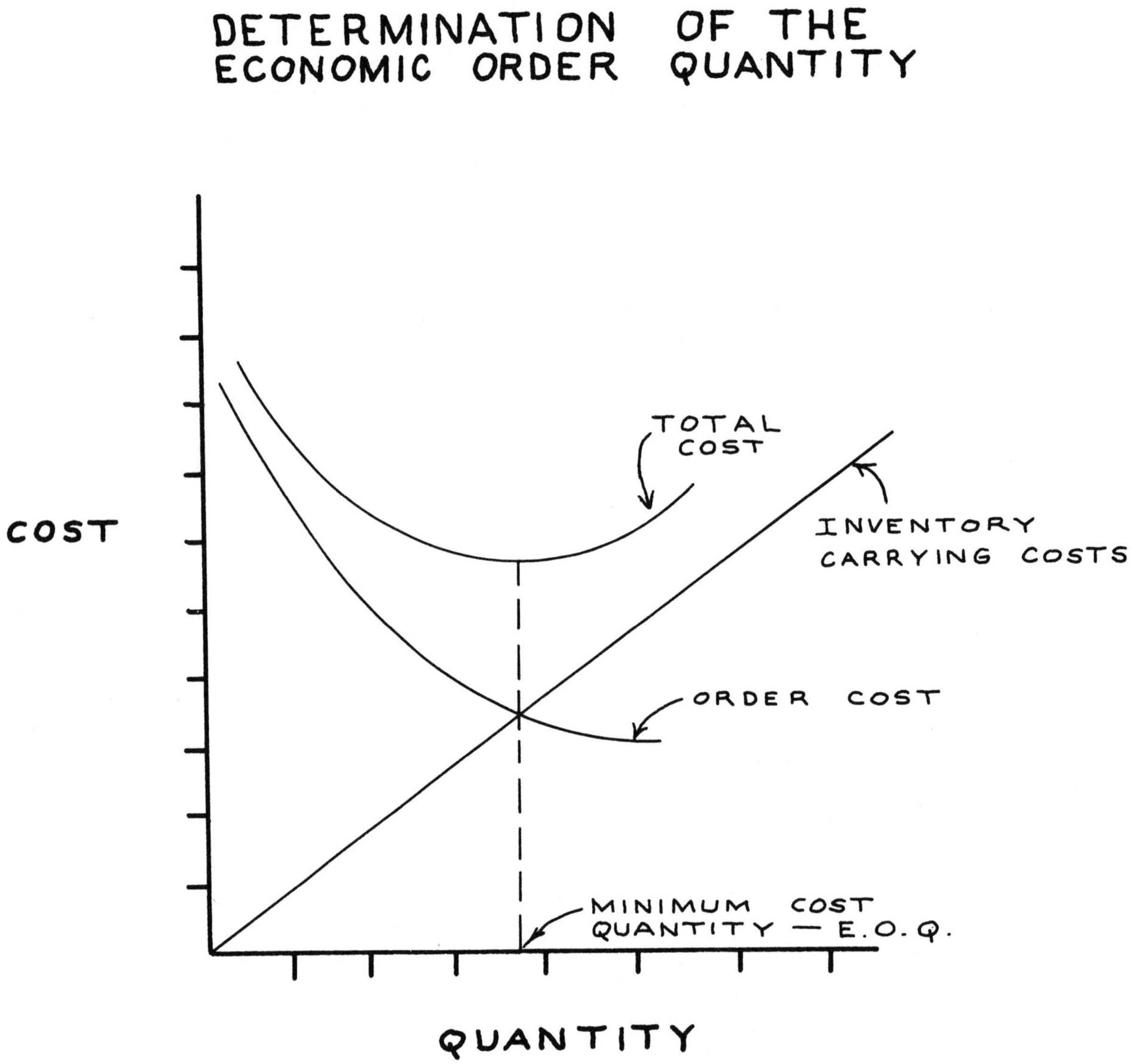

Exhibit 11-2

changes is unnecessarily time consuming. It is much easier to convert the formula into a table, nomograph or slide rule that can be either man- or computer-read. Although it is difficult to determine the cost of ordering and carrying an inventory, the nature of the formula is such that minor variations from true costs tend to even themselves out and give results that are quite useable.

In the Economic Order Quantity method of determining the amount to buy, different factors can change the results completely. If a price change is expected soon, if there are any special storage, depreciation or obsolescence problems and if there is a

Example Showing E.O.Q. with Price Breaks

1. Problem Conditions (Buying Economic Lot Quantities)

   Annual Usage = 500 × 12 = 6,000 pcs.
   Cost to Own = 20%
   Cost to Buy = $15/order
   Unit Cost 600 pcs. = $2.50/ea.

2. Assume Following Price Breaks: 500 pcs. @$2.50/ea.
   1,000 pcs. @$2.40/ea.
   2,000 pcs. @$2.38/ea.

Order Quantity	Unit Cost	Order Value	Average Inventory	Cost to Own 20%	Cost to Buy $15/Order
600	$2.50/ea.	$1,500	$ 750	$150	$150 (10/yr.)
1,000	$2.40/ea.	$2,400	$1,200	$240	$ 90 (6/yr.)
2,000	$2.38/ea.	$4,760	$2,380	$476	$ 37 (2½ yr.)

	1000 pcs.		2000 pcs.
Additional Cost to Own	$ 90		$236
Reduction in Cost to Buy	$ 60		$ 53
Net Saving (Loss)	($ 30)		($183)
Plus Annual Saving 6,000/yr. × $.10	$600	6,000/yr. × $.02	$120
Total Annual Net Saving (Loss)	$570		($ 63)

Illustration shows saving of $570 by going to 1,000, but loss of $63 in taking the next step to 2,000 pcs.

probable change in the usage picture, Economic Order Quantity should be a guide to the decision of the inventory control section rather than the decision itself.

A further point of consideration, and a very important one, is the matter of quantity discounts. In our problem (page 197) the price is set at $2.50/each and our E.O.Q. on this basis is 600 pieces. Should the price be reduced to $2.40 each for a quantity of 1,000 pieces and $2.38 each for 2,000 pieces, we would have this additional factor to consider in determining how much to order. First considering the 1,000-piece quantity, we would have an additional cost to own of $90 but an annual saving in cost of $600 on the purchase price. We should obviously take this step. Next, considering the 2,000 quantity, we would have an added cost to own of $236 over the 1,000 quantities but an additional saving of only $120 on the purchase price. This step would not be economical and we should order only the 1,000 pieces. It is interesting to note that for an additional average investment of under $500 (400/2 × $2.40) an annual saving of over $500 can be realized. In effect the added inventory is being obtained at no cost!

MORE SOPHISTICATED SYSTEMS UTILIZING FORECASTING TECHNIQUES

The inventory management systems previously listed require a good estimate of monthly sales demand. If this data does not exist it must be obtained somehow. It may be necessary to key-punch sales invoices to obtain an accurate product-by-product sales profile.

Relying on product-knowledgeable individuals to provide demand data from their experience leads to trouble. These estimates are usually too inaccurate for inventory management purposes. Some statistical method is needed, even if this means obtaining the last two months of sales and dividing by two. If the data doesn't exist, get it. And in obtaining sales data, it must be by the lowest stockkeeping unit.

The goal in developing more sophisticated systems is to obtain order points based on statistical forecasts of sales demand and order quantities which are based either on economic order quantities (if they are desirable) or on time supplies which are based on a desired level of inventory turnover.

The Order Point is defined as being equal to the demand during lead time and review time plus safety stock. Safety stock is needed to cover three possible occurrences:

(1) Forecast errors. See Exhibit 11-3.

(2) Lead time errors.

(3) The variability of demand. See Exhibit 11-4.

To illustrate how the order point is arrived at, consider the following example:

Assume that demand = 100 units/mo., and that lead time = 6 weeks. In the absence of a mechanical method of determining safety stock, assume S. S. = 1 month's supply or 100 units.

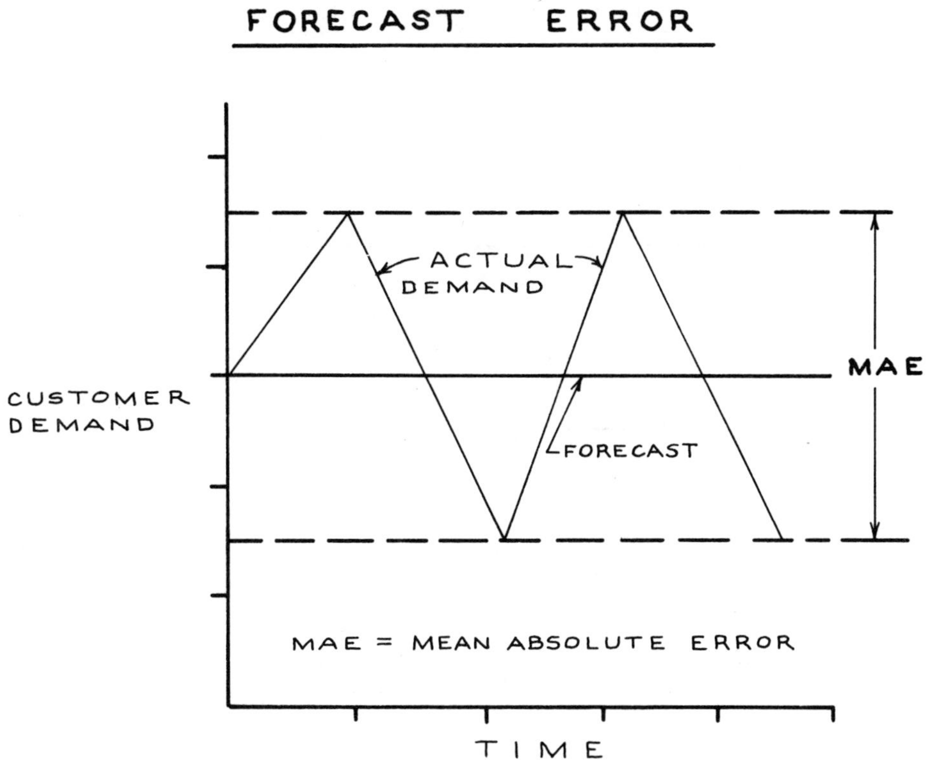

Forecasts may be expected to be too high half the time and too low the other half. Forecast systems measure this error through analysis of demand history and provide for it with safety stock. The greater the MAE is for an item, the higher the safety stock will be.

Exhibit 11-3

Order Point = Demand During L. T. + S. S.
O. P. = 150 + 100
O. P. = 250 units

Safety stock is the most difficult part of the equation to determine. Importantly, the proper determination of safety stock is crucial to the inventory management system's ability to provide a desired level of customer service at minimum investment in inventories. It may be seen, then, that the more accurate the forecast, the less safety stock required. And, the less safety stock you have in your system, the lower your overall inventory level will be. It is for this reason that the remaining pages of this chapter center on Forecasting.

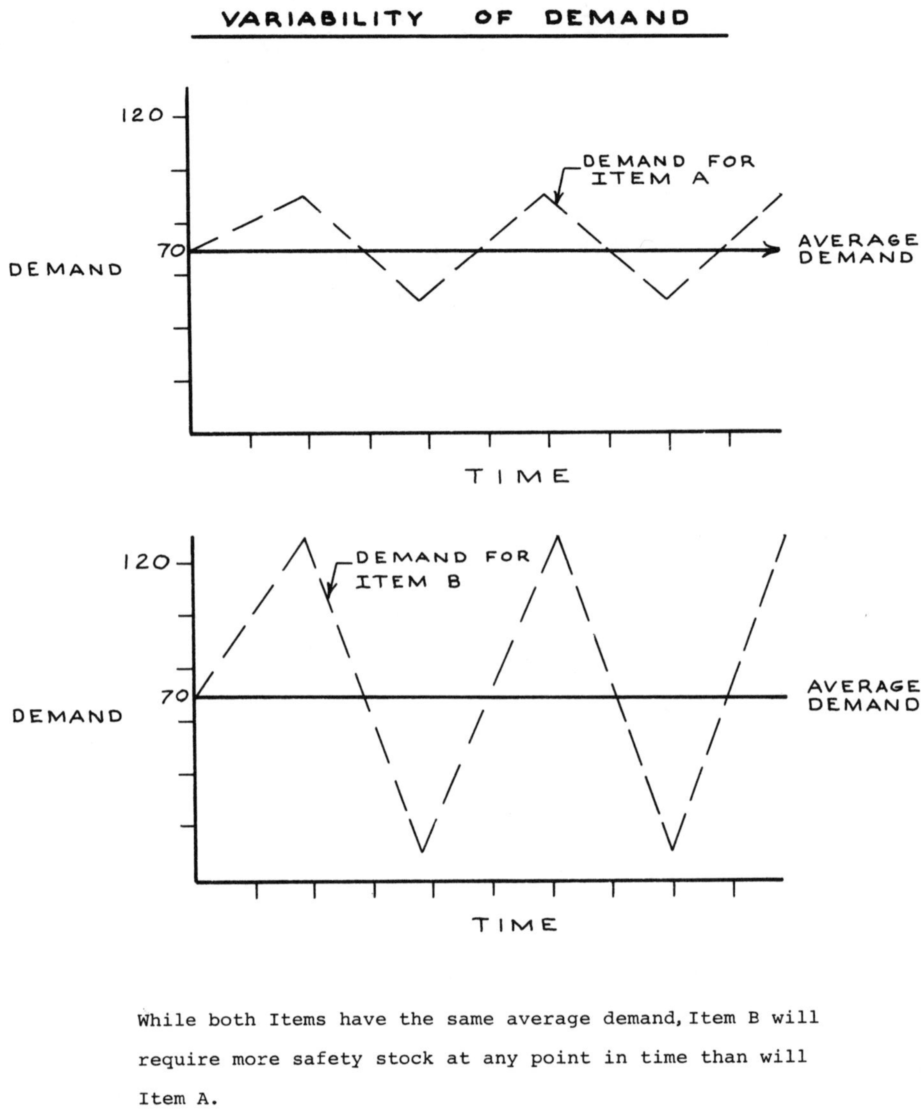

Exhibit 11-4

RESPONSIBILITY FOR FORECASTING

The more sophisticated inventory management systems rely heavily on sales forecast methods which provide fairly accurate average monthly sales values. Forecasting has become a highly specialized scientific field. Several universities now issue doctorates in the subject.

In discussing "forecasts," a distinction should be made between that which is ordinarily generated by Marketing departments and the kind of forecasts which are needed for inventory management. Marketing department forecasts are usually presented by product grouping rather than by the lowest stockkeeping unit as required for inventory management purposes. Marketing forecasts of this sort serve well for budget and sales management purposes, but they are not suitable as input for inventory management systems.

Practically speaking, the Production Planning department, or if a wholesaling firm, the Purchasing department, is usually the source of sales forecasts at the lowest stockkeeping unit level. Interestingly, consolidating and working backwards from the latter forecasts could result in an accurate forecast at the product group level. However, this is rarely done. Unfortunately, the communication gap between marketing and production/purchasing is very broad.

One way in which marketing forecasts are sometimes very far from the mark are where large quantities of return goods are involved. Marketing tends to overlook returns in their calculating, but warehouse personnel are very much aware of them.

The point should be made that accurate sales forecasts can only be generated if all departments cooperate and communicate with each other. Marketing should provide advance information on special promotions and the anticipated sales which they are expected to generate. Sales can indicate when local sales efforts and pressures will affect the forecast. Applying this sort of intelligence to the basic, low-level, statistical forecast can provide a very accurate, detailed picture. By exploding these forecasts backwards, raw material and packaging material requirements can be determined.

One may ask, does forecasting make enough of a difference to make it worth the bother? In one large consumer products company, studies demonstrated that each 1% improvement in the forecast error resulted in a $300,000 decrease in inventory investment. Basically, this meant reductions in safety stock levels. Prior to introducing statistical forecasting, the company had a $10 million inventory to support a 98% customer service level. A highly competent and knowledgeable production manager produced estimated forecasts. After a formal forecasting system was implemented, the company was able to maintain the same 98% customer service level with only $7.5 million in inventory.

Even the simplest form of statistical forecasting is far superior to guessing. Merely setting next month's sales (except where seasonality is involved) equal to the average of the last three months' sales would be better than having the most knowledgeable product manager attempt to estimate sales. The evidence overwhelmingly supports a continuous, statistical approach in arriving at sales forecasts. There is absolutely no doubt that in the absence of a forecast system excess safety stock will have to be maintained to provide a given level of customer service.

METHODS OF FORECASTING

Forecasting, as the term is used in this chapter, refers to the objective measure of historical sales data in order to determine what sales will be in the future. How objective measurement is accomplished is the subject of this section.

MOVING AVERAGES

One of the simplest and least expensive ways to obtain a forecast is to calculate moving averages of historical sales. This would require, for example, taking an average of the last four months of sales to develop a forecast for the next month. Then, each month the oldest month is dropped and the most recent month of sales history is added and a new four-month average calculated. With trend items, a two-month moving average might be appropriate; with more stable items, a six-month moving average could be used.

A moving average method of arriving at a forecast is superior to simply taking last month's sales as representing next month's sales, because factors unknown to the forecaster can cause too great a distortion in the short-run. Moving averages tend to smooth out distortions to the data.

The low cost of obtaining moving averages makes it attractive as a first approach to forecasting. It may be done manually, on time-sharing, through a service bureau or on a company's captive computer. For example, the monthly time-sharing cost to obtain a three-month moving average for 5,000 items would amount to only $100. Further, no expensive programming is required at the start of such an effort. Sophisticated forecast systems require programs which can cost thousands of dollars.

WEIGHTED MOVING AVERAGE METHOD

This method is a step above simple, moving averages in sophistication. It is too time consuming to be done manually, but it can be done on time-sharing. It is a good approach for firms who do not wish to spend the large amounts of money which more sophisticated systems require to initiate and run.

The weighted moving average method would use, for example, a three-month horizon where demand three months ago would be weighted at 20%; demand two months ago at 30% and last month's demand at 50%. The table below illustrates the method:

Percent ×	Successive Monthly Sales =	Monthly Weights
20%	80 Units	16 Units
30%	90 Units	27 Units
50%	100 Units	50 Units

Total Forecast for Next Month = 93 Units

EXPONENTIAL SMOOTHING

This is the most practical and economically available system of the more sophisticated approaches to forecasting.

Exponential smoothing is a more advanced form of the weighted average technique. To simplify as much as possible, the forecast obtained in this manner is equal to some fraction of last month's demand plus some exponentially smaller fraction of the previous month's demand and so on. The process usually extends back for 12 months, with greater weight given to the most recent months.

This method gives a very good operational forecast. Its limitation, which is really a limitation of all of the forecast methods discussed previously, is that it cannot be used to extrapolate more than three months into the future.

EXAMPLES OF RESULTS WHICH COMPANIES ACHIEVED IN USING FORECASTS

- Prior to analyzing its inventory situation, a plumbing supply wholesaler operated with a $2 million inventory, turned its inventory 1.67 times per year and had a customer service level of 68%.

  A time supply system based on average monthly sales taken from existing ledger cards was installed. Within one year, turnover improved to 2.43, representing roughly a 30% reduction in inventory.

  During this first year, an A-B-C analysis of the firm's inventory was performed. An exponential smoothing system was applied to the "A" items (representing the top 20% of sales) and an average demand system based on a three-month moving average was used for the "B" and "C" items. After 18 months, turnover was up to 3.00.

  Next, the "B" items were included in the exponential smoothing system with the "A" items. At the end of two years, the company was turning its inventory at the rate of four turns per year. Customer service increased to 88%, while the company operated with less than $1 million in inventory.

- A floor covering company devoted two and a half months to installing an exponential smoothing system to cover all of its 3,000 items. Within 12 months, inventory turns increased from 4.00 to 7.50; inventory value decreased from $1 million to approximately $600,000. Customer service, which was high to begin with, remains the same.

- A simple manual forecast system developed for a small health and beauty aids wholesaler/retailer, resulted in increasing inventory turns from 2 to 3.5 per year.

  The company supplied six branch, retail outlets from a single warehouse. Each branch manager was asked to list the top 100 products in his line and to also provide an estimate of monthly sales for each item.

  A time supply system was then developed for these "A" items showing order points and order quantities. "B" and "C" items were handled as before. The improvement in turnover was achieved in part by having the branches order in relatively small time supplies (lead time was only two and a half days). This had the effect of smoothing out demand at the warehouse level and this resulted in lower safety-stock requirements.

- A home furnishings wholesaler specializing in flatware and dishes, increased turnover from 1.6 to 3.5 per year. This was accomplished by closing out many slow moving items.

  In manually analyzing its inventory, the company found that many items did not move for from 24 to 36 months. It developed that inventory requirements had been based on Marketing department estimates. These estimates were prepared by three separate managers, each of whom sold the entire product line. Each manager added his own safety-stock factor in addition to inflating his sales estimates. The company did not monitor actual sales performance against the forecasts.

- Having a perpetual inventory card system, an assembler and wholesaler of imported light fixtures had experienced 1.2 turns per year on an inventory of $2 million.

  The firm was found to have a problem in its unsophisticated ordering strategies. They simply ordered when they thought they needed something and they ordered what they thought was a good amount.

  Utilizing a manual time supply system, which showed appropriate order points and order quantities, inventories were reduced by $400,000 in nine months.

SUMMARY—FORECASTING CONSIDERATIONS

1. Decide on forecast interval and frequency.
2. See what data is available:
   a. Check to see if each period has the same number of working days.
   b. Remove effect of promotions, commissions.
3. Choose and test a forecasting method. Calculate the effect of smaller forecast error.
4. Measure the forecast error quantitatively.
5. Use formalized forecast monitoring.
6. Make proper use of each updated forecast.

12

Coupling Good Inventory Turnover with Better Customer Service

Being able to maximize customer service while simultaneously achieving improved inventory turnover requires good inventory performance tools. This chapter provides a clear outline of what reports are needed and, specifically, how to design and implement them.

A. CUSTOMER SERVICE AND INVENTORY TURNOVER

Inventories exist for one purpose—to satisfy customer demand. But, every businessman knows that some constraints must be placed on inventory levels, and that it is not possible to have an infinite supply of every item. The question is, just what kind of constraints should be placed on inventories?

This leads into the two primary inventory management goals: High customer service levels and maximizing inventory turnover. Unfortunately, these goals are always in conflict with one another. If your goal is to satisfy every customer order as it is received, very high levels of inventory will be required. Many studies have been made which demonstrate the fact that inventories increase exponentially as attempts are made to achieve service levels approaching 100% (see Exhibit 12-1). This is the case because extra inventories must be maintained to handle infrequent but unusually large customer demands. Each company must make its own decision as to the level of

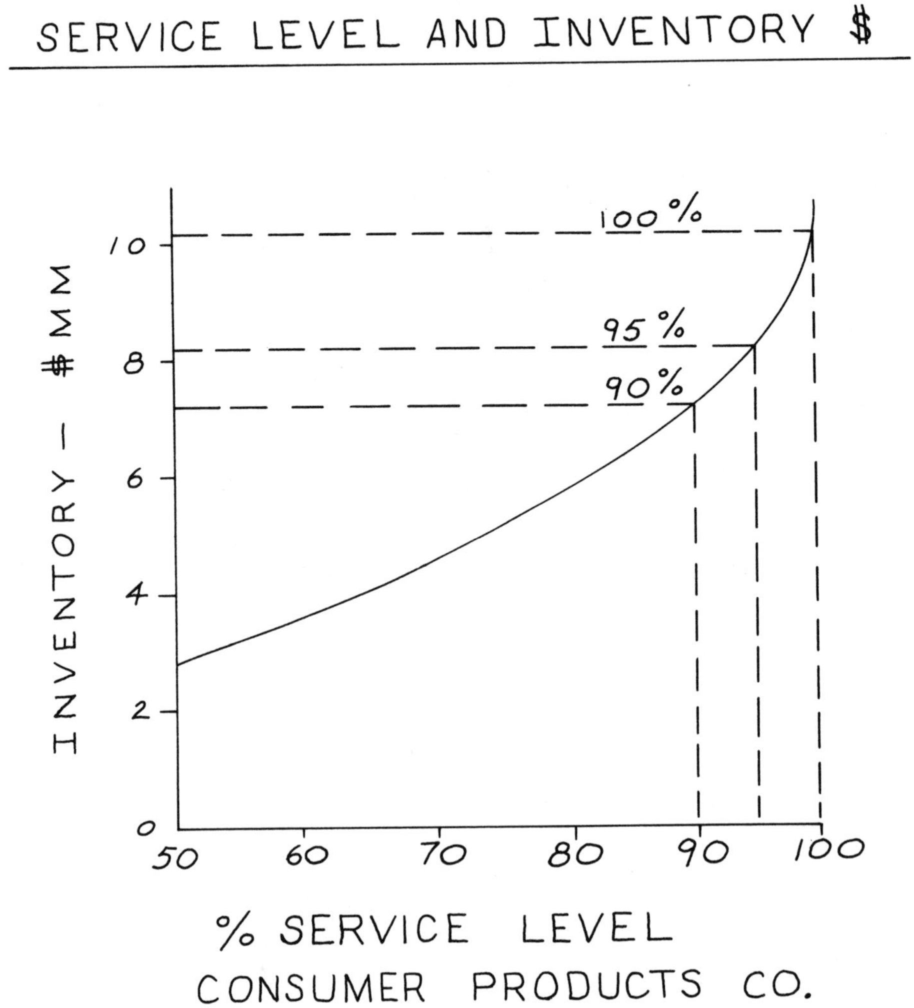

Exhibit 12-1

customer service which it wants to achieve. The job then becomes one of providing the minimum inventory necessary to meet that customer service objective.

There are several definitions of customer service. One is the percentage of dollar demand of customer orders satisfied on time. On time could be within a day, a week, a month or any time period that you wish to choose. Other criteria may be the percent of line items satisfied on time, the percent of units satisfied on time or the percent of orders filled on time. Percent of complete orders filled on time may serve as an important Sales department measure of customer service, but it is not the best measure of inventory management system performance. This may be seen in the fact that a stockout of a single popular item can result in a large number of orders being shipped incomplete.

Inventory turnover is usually calculated thus:

$$\frac{\text{Total Annual Sales at Cost}}{\text{Average Monthly Inventory Value at Cost}}$$

Retail businesses often measure turnover in the following way:

$$\frac{\text{Total Annual Retail Sales}}{\text{Average Monthly Inventory Value at Selling Price}}$$

Whatever definition is used, it is the actual measure of physical turnover of inventory. So that, if you are achieving six turns per year, you actually have two months' supply on hand all the time (12 mos. ÷ 6 turns = 2 mos. supply).

Turning now to the real problem, you want to first determine what the customer service level objective should be, and then you need to calulate the amount of safety stock required for each item to achieve that level of service. Turnover statistics should be developed by item in addition to the total inventory. This permits evaluation of each item's contribution to the overall turnover objective and facilitates corrective action where necessary. Perhaps, for example, if management were given a list of slow and non-movers, they would authorize deleting them from the product line.

The importance of customer service vis-a-vis turnover as a measure of management performance, depends on the nature of the marketplace. In highly competitive industries, customer service would be more critical than turnover. However, even under highly competitive situations, it is possible to compare your turnover performance with the inventory turns which others in your industry are experiencing. Dunn and Bradstreet's annual business performance indexes show the median, lowest quarter mean and highest quarter mean turnover statistics for all major wholesale, manufacturing and service industries.

> *CASE IN POINT:* A consumer products manufacturer, operating out of four warehouses, used its service level report to evaluate item availability as well as overall service level. Management felt that total service level was too low—90%. The service level report showed that virtually all items being back-ordered were either special-priced merchandise which were manufactured to customer order and not carried in inventory, or "seconds" which were a by-product of the manufacturing process. Management decided to eliminate these two types of items from the service level calculation. Service level then rose to 98%. This number more accurately represented the company's ability to support its sales objectives.

Some firms might also find it useful to show the number of units ordered and back-ordered.

1. The % shipped column (see Exhibit 12-2) shows the percentage dollar value shipped against the total dollar value ordered for each item. This is a good reference number which provides fast perspective to the back-ordered amounts.
2. The last column shows the number of times that an item was deleted from an order. If a 2 is displayed, that means that the item was deleted from two orders. (Another Service Level Report is shown in Exhibit 12-4.)

V-000 ** JOBBERS ** WEEKLY SERVICE LEVEL REPORT WEEK ENDING 06/16/ 4202021-01

PRODUCT	DESCRIPTION		$ ORDERED	$ BACK-ORDER	$ SHIPPED	% SHIPPED	# TIMES BACK-ORDER
9610-300	MURIEL	CORONELLA	282.64		282.64	100.0	
	NEW LONG SIZE						
9610-379	MURIEL	CORONELLA 5/30	565.28	565.28		.0	1
9610-420	MURIEL	CORONELLA	565.28		565.28	100.0	
	NEW LONG SIZE						
9610-435	MURIEL	CORONELLA 5/28				.0	
	NEW LONG SIZE						
9610-440	MURIEL	CORONELLA 5/29				.0	
9610-442	MURIEL	CORONELLA 5/30				.0	2
9610-816	MURIEL	CORONELLA VEND	459.29	459.29		.0	3
	NEW LONG SIZE						
9720-300	MURIEL	MINT & MENTHOL	24.97		24.97	100.0	
	AIR TIPS						
9720-379	MURIEL MINT	& MENTHOL 5/30				.0	
4015-030	FACTORY SMKRS	NO 9	50.00		50.00	100.0	
4015-300	FACTORY SMKRS	NO 9	100.00		100.00	100.0	
4210-300	MURIEL	PERFECTO				.0	
5020-030	FACTORY SMKRS	NO 9				.0	
5020-300	FACTORY SMKRS	NO 9				.0	
5081-030	H & M	STRAIGHT				.0	
5093-030	LANE	FACTORY SECDS				.0	
5610-351	DUTCH MASTERS	PERFECTOS 5/63				.0	
5610-824	DUTCH MASTERS	PERFECTOS	35.51		35.51	100.0	
5610-825	DUTCH MASTERS	PERFECTOS	35.51		35.51	100.0	
5610-914	DUTCH MASTERS	PERFECTO 5/67				.0	
5640-030	DUTCH MASTERS	PRESIDENTS	337.55		337.55	100.0	
5640-406	DUTCH MASTERS	PRESIDENT 5/75				.0	
5660-914	DUTCH MASTERS	SPEC BLT 5/67				.0	
5900-300	DUTCH MASTERS	CADETS TIP	73.66		73.66	100.0	
5900-357	DUTCH MASTERS	CAD TIPS 5/35				.0	
5900-404	DUTCH MASTERS	CADET TIP 5/37				.0	
6050-030	ADLON	PERFECTO				.0	
6050-300	ADLON	PERFECTO				.0	
6050-336	ADLON	PERFECTO 5/57				.0	
8010-030	EL PRODUCTO	BOUQUETS	1,499.50		1,499.50	100.0	
8010-125	EL PRODUCTO	BOUQUETS				.0	
	IMPRINTED						
	WITH CAUTION NOTICE						
8010-300	EL PRODUCTO	BOUQUETS	1,349.55		1,349.55	100.0	
8010-374	EL PRODUCTO	BOUQUETS 5/65	149.95		149.95	100.0	
8010-824	EL PRODUCTO	BOUQUETS				.0	
8010-825	EL PRODUCTO	BOUQUETS	35.99		35.99	100.0	
8010-914	EL PRODUCTO	BOUQUET 5/67				.0	
8170-030	LOVERA	PERFECTO	2,196.40		2,196.40	100.0	
8170-300	LOVERA	PERFECTO	686.38		686.38	100.0	
8180-030	LOVERA	LONG STAR	422.33		422.33	100.0	
8180-300	LOVERA	LONG STAR	563.10		563.10	100.0	
8240-030	EL PRODUCTO	BLUNTS	446.63		446.63	100.0	
8240-125	EL PRODUCTO	BLUNTS				.0	
	IMPRINTED						
	WITH CAUTION NOTICE						
8310-030	EL PRODUCTO	DIAMOND TIPS	59.81		59.81	100.0	

Exhibit 12-2

```
V-000          ** JOBBERS **     WEEKLY SERVICE LEVEL SUMMARY     WEEK ENDING 06/16/     4202021-02

     TOTAL NUMBER OF ORDERS PROCESSED                        1,166
     TOTAL NUMBER OF ORDERS SHIPPED INCOMPLETE                 184
     TOTAL NUMBER OF ORDERS SHIPPED COMPLETE                   982
     PERCENT OF ORDERS SHIPPED COMPLETE                       84.0


     TOTAL DOLLAR VALUE ORDERED                      $1,618,405.57
     TOTAL DOLLAR VALUE BACK-ORDERED                    $52,224.77
     TOTAL DOLLAR VALUE SHIPPED                      $1,566,180.86
     PERCENT DOLLAR VALUE SHIPPED                             96.7
```

Exhibit 12-3

SERVICE LEVEL EVALUATION REPORT DATE 03/31/

PART	DESCRIPTION	QTY ORDERED	QTY SHIPPED	$ ORDERED	$ SHIPPED	SERVICE LEVEL	THEO INV $	THEO INV-TURN	OLD O.P.	NEW O.P.
4051	GE DISH			.00	.00	.00	650.68	1.27	57	62
4053	MOTOR			12.73	12.73	100.00	210.96	4.92	21	22
4055	MOTOR	152	77	1614.24	817.74	50.65	2114.62	5.05	219	252
4057	GE AW	32	32	358.40	358.40	100.00	1663.33	5.66	198	192
4059	MOTOR	30	30	364.80	364.80	100.00	519.02	9.22	70	64
4061	MOTOR	53	53	656.67	656.67	100.00	1164.61	7.20	145	129
4063	UNIVERSAL DR	353	353	4034.79	4034.79	100.00	3913.54	12.71	607	566
4065	MOTOR	10	10	110.90	110.90	100.00	493.03	5.88	60	59
4067	MOTOR	137	137	1633.04	1633.04	100.00	1492.39	8.50	152	181
4069	MOTOR	152	152	1673.52	1673.52	100.00	2809.37	7.51	400	354
4071	MOTOR	3	3	35.31	35.31	100.00	374.17	4.52	41	40
4073	MOTOR	1	1	11.59	11.59	100.00	56.66	4.66	6	7
4079	MOTOR	6	6	78.06	78.06	100.00	99.86	4.69	9	10
4081	MOTOR	2	2	33.18	33.18	100.00	136.42	6.12	12	11
4083	MOTOR	35	35	573.65	573.65	100.00	562.43	5.94	33	45
4087	MOTOR	54	54	858.60	858.60	100.00	1377.28	7.21	135	119
4089	MOTOR	76	76	1127.84	1127.84	100.00	2306.18	7.11	234	212
4091	MOTOR	26	26	343.98	343.98	100.00	1392.84	6.22	143	140
4092	D & M DISH	242	167	3032.26	2092.51	69.00	2843.72	7.51	250	314
4095	MOTOR	96	96	1143.36	1143.36	100.00	1172.20	6.02	99	130
4096	MOTOR	51	51	554.37	554.37	100.00	1354.34	6.12	183	165
4097	MOTOR	28	28	356.44	356.44	100.00	1102.10	5.68	122	112
4098	MOTOR	48	48	630.72	630.72	100.00	956.05	8.34	117	104
4099	HOTPOINT DR	9	9	101.34	101.34	100.00	262.23	10.25	32	36
4101	HOTPOINT DR	5	5	56.10	56.10	100.00	181.50	6.75	23	22
4103	MOTOR	9	4	101.88	45.28	44.44	257.37	6.54	33	31
4105	MOTOR	20	20	214.60	214.60	100.00	324.56	9.44	49	46
4108	MOTOR	29	29	298.99	298.99	100.00	335.83	7.18	41	45
4109	MOTOR	78	78	840.06	840.06	100.00	899.51	6.89	97	113
43RA5	OVEN CONTACT	25	24	3.20	3.07	95.93	1.63	5.27	17	17
480	BUSHING	598	502	137.54	115.46	83.94	691.40	5.97	3364	3179
4801B	INSERT MED	2340	1246	655.20	348.88	53.24	574.77	10.71	2242	2258
4801C	INSERT LARGE	368	274	103.04	76.72	74.45	128.32	7.43	632	632
480017	DRIER	32	32	55.04	55.04	100.00	138.87	5.87	118	106
5_440	BELT	105	105	104.05	104.05	100.00	97.86	12.56	156	163
500903	CASE	999	207	309.69	64.17	20.72	807.22	3.76	4905	4739
50100C	RESTRING KIT	3		3.18	.00	.00	3.36	5.30	3	5
50210	RESIST WIRE			.00	.00	.00	14.47	3.95	8	8
5036	36 CORD RANGE	178	72	206.48	83.52	57.44	155.28	.43	112	255
5048	48 CORD RANGE	3	3	4.20	4.20	100.00	146.00	.00	105	85

Courtesy: Schoolman Consultants Inc.

Exhibit 12-4

Exhibit 12-3 is a Weekly Service Level Summary Report. It shows the total number of orders processed and the total shipped complete and incomplete. A percentage figure of orders shipped complete is then indicated. The report shows another measure of customer service; i.e., percentage of dollar value shipped. This summary by itself serves as a useful top management information feedback report.

Exhibit 12-5 presents one page from an actual Inventory Aging and Turnover Report. Note that all items in the report show zero turnover. This results from the fact that these items had no sales for the last six months.

CASE IN POINT: A restaurant supplies distributor felt that additional warehouse space was needed. The product line had expanded over the years along with total sales. Management wanted an efficient, smooth-operating warehouse. A consulting firm was called in to develop a new layout and determine how much additional space would be required. As a first step, the consultants requested that the firm's Accounting department prepare an aging report similar to that shown in Exhibit 12-5. It was the first time that such a report was prepared by the company. The report showed that 22% of the line items being carried in inventory had not had a single sale in from two to four years! Spacewise, these items were accounting for 33% of the floor area. Management decided to eliminate most of the product line. In so doing, the company avoided new construction and saved $150,000 in the process.

Even the best inventory management system cannot eliminate all of the high supply period items. Some will always slip through. Aging analysis is needed on a continued basis for monitoring purposes.

Shown below is a computer program specifications sheet covering the inventory aging and turnover analysis report shown in Exhibit 12-5.

B. COMPUTER PROGRAM SPECIFICATIONS FOR INVENTORY AGING AND TURNOVER ANALYSIS REPORT

This is a report showing the months of supply of inventory at present and projected at six intervals into the future for each product-package. A sample page is enclosed. This is run in conjunction with the Usage by Value Report.

Input

1. The same data used by Usage by Value.
2. The item count dollar annual sales and dollar value of inventory determined for the Usage by Value Report.

Calculations

1. Calculate the total months of supply of inventory by item.

$$\text{Months of Supply} = \frac{\$\ \text{inventory}}{\$\ \text{annual sales}} \times 12$$

INVENTORY AGING AND TURNOVER ANALYSIS MAR 15,

PART NUMBER	PROD	ITEM COUNT	UNIT COST	DOLLAR ANNUAL SALES	DOLLAR INVENTORY	CUM DOLLAR INVENTORY	MONTHS OF INVENT	TURN OVER	6 MOS	12 MOS	18 MOS	24 MOS	36 MOS	60 MOS
999E003		4045	.2170	0	7000	7000	999	.00	7000	7000	7000	7000	7000	7000
999F003		4046	.3100	0	2365	9365	999	.00	2365	2355	2365	2365	2365	2365
999F008		4047	.0873	0	1349	10714	999	.00	1349	1349	1349	1349	1349	1349
286P503		4048	.1447	0	1287	11996	999	.00	1282	1282	1282	1282	1282	1282
999F034		4049	.1228	0	1216	13212	999	.00	1216	1215	1215	1216	1216	1216
897B003		4050	.0350	0	1129	14341	999	.00	1129	1129	1129	1129	1129	1129
925W691		4051	8.1000	0	1110	15451	999	.00	1110	1110	1110	1110.	1110	1110
999D013		4052	.1200	0	945	16396	999	.00	945	945	945	945	945	945
287B002		5053	.1640	0	943	17339	999	.00	943	943	943	943	943	943
5911106		4054	1.1682	0	930	18269	999	.00	930	930	930	930	930	930
5911093		4055	3.4900	0	873	19142	999	.00	873	873	873	873	873	873
728C500		4056	.4394	0	801	19943	999	.00	801	801	801	801	801	801
999K001		4057	.0106	0	742	20685	999	.00	742	742	742	742	742	742
286P501		4058	.1471	0	673	21358	999	.00	673	673	673	573	673	673
999F037		4059	.1085	0	622	21980	999	.00	622	522	622	522	622	622
286P502		4060	.1421	0	607	22587	999	.00	607	507	607	507	607	607
336B103		4061	.0665	0	579	23166	999	.00	579	579	579	579	579	579
999F002		4062	.0701	0	506	23672	999	.00	506	506	506	506	506	506
728C526		4063	.6240	0	496	24168	999	.00	496	496	496	496	496	496
999F023		4064	.0607	0	433	24601	999	.00	433	433	433	433	433	433
999D015		4065	.1650	0	413	25014	999	.00	413	413	413	413	413	413
999K004		4066	.0214	0	411	25425	999	.00	411	411	411	411	411	411
351E502		4067	.0175	0	406	25831	999	.00	406	406	406	406	406	406
999G002		4068	.0525	0	393	26224	999	.00	393	393	393	393	393	393
999B008		4069	.0398	0	376	26600	999	.00	376	376	376	376	376	376
999B007		4070	.0398	0	376	26976	999	.00	376	376	376	376	376	376
W36741R		3531	.5750	3	368	27344	999	.00	366	355	363	352	359	353
999D006		4071	.0600	0	366	27710	999	.00	366	356	366	356	366	366
999F026		4072	.1334	0	352	28062	999	.00	352	352	352	352	352	352
999A009		4073	.2660	0	348	28410	999	.00	348	348	348	348	348	348
591C504		4074	.1800	0	346	28756	999	.00	346	346	345	346	346	346
591C501		4075	.1800	0	343	29099	999	.00	343	343	343	343	343	343
WB6740G		3532	.5750	3	342	29441	999	.00	340	339	337	336	333	327
999F020		4076	.0226	0	339	29780	999	.00	339	339	339	339	339	339
337B105		4077	.1055	0	338	30118	999	.00	338	338	338	338	338	338
999A031		4078	.1500	0	337	30455	999	.00	337	337	337	337	337	337
999C004		4079	.0130	0	326	30781	999	.00	326	326	326	326	326	326
163B013		4080	.0189	0	324	31105	999	.00	324	324	324	324	324	324
999B030		4081	.0472	0	307	31412	999	.00	307	307	307	307	307	307
999B029		4082	.0466	0	303	31715	999	.00	303	303	303	303	303	303
999C010		4083	.0115	0	299	32014	999	.00	299	299	299	299	299	299
775E003		4084	.0035	0	298	32312	999	.00	298	298	298	298	298	298
999F021		4085	.0193	0	290	32602	999	.00	290	290	290	290	290	290

Courtesy: Schoolman Consultants Inc.

Exhibit 12-5

2. Calculate theoretical turnover by item

$$T/O = \text{annual sale} \div \$\ \text{inventory}$$

3. Calculate the dollar value of inventory that will be left after 6, 12, 18, 24, 36 and 60 months by item.

   a) \$ inventory> 6 mos. =
   $$\$\ \text{total inventory} - \frac{\text{annual sales}}{2}$$

   b) \$ inventory> 12 mos. =
   \$ total inventory — annual sales

   c) \$ inventory> 18 mos. =
   \$ total inventory — 1.5 × annual sales

   d) etc.

C. DISTRIBUTION BY VALUE

Distribution by value analyses are also referred to as A-B-C analyses or usage by value analyses. They are all used to show a ranking of items by their contribution to total sales.

A firm's ordering strategies and customer service level criteria are often set by whether an item is a slow, medium or fast mover. It might, for example, be desirable to establish higher customer service level criteria on fast movers than on slow movers. You might want to review the activity of key, fast moving items more frequently than slow ones. Some purchasing managers establish slightly higher order quantities on the slower moving items in order to reduce the workload on personnel, while not affecting total turnover or customer service very much.

> *CASE IN POINT:* A paper products distributor found from a distribution by value analysis that the bottom 30% of items carried in inventory accounted for only 2% of sales.
>
> The Purchasing department, consisting of seven persons, was devoting an equal amount of attention to maintaining adequate inventories on all items. Weekly inventory status review was conducted on every item. A "Bin Reserve" procedure was adopted for slow moving items, whereby warehouse personnel would alert purchasing when supplies were running low.
>
> Purchasing provided the warehouse with reorder tickets for the slow movers. The tickets were attached to the last batch or carton in a lot and sent to Purchasing when supplies dwindled to that level. Purchasing was relieved of performing a tedious task and was able to reduce its staff by two persons. Normal attrition was allowed to make the reduction in personnel.

Distribution by value reports should be generated periodically on a continuing basis. The profiles of inventories change from period to period, from year to year, and the inventory manager needs to know what these changes are so that he may alter his ordering strategies.

USAGE BY VALUE (ABC) MAR 15,

PART NUMBER	ITEM COUNT	CUM% ITEM COUNT	COST	UNIT SALES	DOLLAR ANNUAL SALES	CUM DOLLAR SALES	CUM% DOLLAR SALES	UNITS INVENT	DOLLAR INVENT	CUM DOLLAR INVENT	CUM% DOLLARS INVENT
561-003	1	.02	.5157	22868	11793	11793	1.44	3792	1956	1956	.67
381X001	2	.04	1.6987	6869	11668	23461	2.87	2252	3825	5781	1.97
561-002	3	.06	.3113	33962	10572	34033	4.15	2736	852	6633	2.26
286P001	4	.08	.1438	69762	10023	44055	5.38	13	2	6635	2.26
921S009	5	.10	.6534	13992	9142	53198	6.50	94	61	6696	2.29
433-001	6	.12	.5330	16499	8794	61992	7.57	60	32	6728	2.30
921S005	7	.14	.5940	14523	8627	70619	8.63	9	5	6733	2.30
281R001	8	.16	.1133	62285	7057	777676	9.49	7	1	6734	2.30
561-001	9	.17	.1368	48029	6570	84246	10.29	10566	1445	8179	2.79
282R001	10	.19	.1200	52644	6317	90563	11.06	6720	806	8985	3.07
351H002	11	.21	.0766	76652	5872	96435	11.78	45450	3481	12466	4.26
352B001	12	.23	.3112	18461	5745	102180	12.48	0	0	12466	4.26
321C001	13	.25	.4621	11763	5436	107616	13.15	1145	529	12995	4.44
351B004	14	.27	.0876	55689	4878	112494	13.74	11054	968	13963	4.77
591E003	15	.29	18.5960	258	4798	117292	14.33	0	0	13963	4.77
287R004	16	.31	.3355	14250	4781	122073	14.91	41	14	13977	4.77
433-002	17	.33	.8350	4948	4132	126205	15.42	2308	1927	15904	5.43
A5116AC	18	.35	.4338	9445	4097	130302	15.92	830	360	16264	5.55
286R003	19	.37	.1700	23976	4076	134378	16.42	4097	696	16960	5.79
250C009	20	.39	.2304	16482	3797	138175	16.88	2164	499	17459	5.96
250B002	21	.41	.2086	18194	3795	141970	17.34	3653	762	18221	6.22
381B002	22	.43	2.6400	1396	3685	145655	17.79	0	0	18221	6.22
R560HA	23	.45	.7150	5130	3668	149323	18.24	0	0	18221	6.22
654T001	24	.47	.4700	7599	3572	152895	18.68	804	378	18599	6.35
5611001	25	.48	.5976	5849	3495	156390	19.11	10	6	18605	6.35
5911009	26	.50	40.5490	85	3447	159837	19.53	0	0	18605	6.35
391X001	27	.52	.2200	15504	3411	163248	19.94	1375	303	18908	6.45
834R001	28	.54	.1150	29082	3344	166592	20.35	792	91	18999	6.49
946M095	29	.56	2.3700	1384	3280	169872	20.75	93	220	19219	6.56
811C013	30	.58	.2350	13275	3120	172992	21.13	0	0	19219	6.56

Courtesy: Schoolman Consultants Inc.

Exhibit 12-6

Exhibit 12-6 shows the best format to use for usage by value analysis. The discussion below follows the bottom line of the report across each of the columns.

1. Part Number.
2. Item Count. This is the fifty-first item ranked in order of its contribution to total sales.
3. Cumulative % Item Count. This shows that the first 51 items represent less than 1% (.99) of the total items carried by the company.
4. Cost. The cost of the fifty-first item.
5. Unit Sales. The number of units of that item sold during the last 12 months. Any time period may be used.
6. Dollar Annual Sales. The dollar value of annual unit sales of that item.
7. Cumulative Dollar Sales. The dollar value of 12 months sales for the first 51 items carried in inventory.
8. Cumulative % Dollar Sales. This shows that the top 51 items account for 27.51% of total dollar sales. Remember that these items represent less than 1% of the total items carried in inventory.

DISTRIBUTION BY VALUE

A	B	C	D	C ÷ D	(D ÷ C) x 52
ITEM COUNT NO.	% OF ITEMS	$ (000) SALES	$ (000) INV.	TURNS PER YEAR	WEEKS OF INV.
17	10%	46,649	4,694	10.0	5.2
84	40%	38,049	4,476	8.5	6.1
126	25%	4,015	585	6.9	7.5
168	25%	1,648	340	4.8,	10.8
TOTAL	100%	90,361	10,095		
	AVERAGE T/O			8.85	
	AVG. WEEKS SUPPLY				5.9

Exhibit 12-7

9. Units in Inventory.
10. Dollar Value of Units in Inventory.
11. Cumulative Value of the first 51 items in inventory.
12. Cumulative %—age of the value of the first 51 items in inventory. The first 51 items account for 9.10% of the total dollar inventory.

Exhibit 12-7 shows a typical Summary Distribution by Value Report. Generated twice each year, it clearly shows turnover performance over time. It also breaks the product line into four categories (sometimes only three breakdowns are made; hence, the term A-B-C analysis).

The table shows that the top 17 items in contribution to sales or 10% of the total items in the product line are experiencing 10.0 turns per year. The bottom 25% of the items are turning only 4.8 times per year. Average turnover for the entire product line is 8.85 and the corresponding weeks of supply is 5.9.

Computer program specifications covering the Usage by Value Report illustrated in Exhibit 12-6 are shown below:

Program Specifications for
Usage by Value Report

This is a report listing each product package in descending order of annual sales at cost.

Shown also is the inventory at any point in time of each product package.

Input

1. 12 months unit sales data for each product package.
2. Cost of each product package.
3. A unit inventory figure for each product package.
4. Product package code for each product package.
5. Product package description.

Calculations

1. Calculate $ annual sales at cost for each product package.
2. Sort in descending order of $ sales.
3. Assign item number to each.
4. Calculate cumulative item count as percent of total item count.
5. Cumulate sales according to ranking.
6. Calculate cumulative sales figure as percent of total sales at cost.
7. Calculate the $ value of the inventory of each product package.
8. Cumulate dollar inventory according to ranking.
9. Calculate cumulative inventory $ figure as percent of total inventory.

PLIMPTON PRESS DIVISION

INVENTORY COMPARISON--INK

Month of February

Fiscal Year 19 /19

INK	Fiscal Year 19	Fiscal Year 19	(Decrease) Increase Over 19
Norwood Plant			
Rotary	6,836	6,879	(43)
Perfector	160	352	(192)
Job	-	-	-
Web	29,537	32,169	(2,632)
Offset	26,862	23,962	2,900
Crabtree	130	883	(753)
Norwood Total	63,525	64,245	(720)
Pinola Plant			
Rotary		942	(942)
Perfector		1,427	(1,427)
Job			
Web	31,847	14,703	17,144
Offset	1,198	3,618	(2,420)
Pinola Total	33,045	20,690	12,355
Total	96,570	84,935	11,635

Exhibit 12-8

PLIMPTON PRESS DIVISION

INVENTORY COMPARISON--SUPPLIES

Month of February

Fiscal Year 19 /19

SUPPLIES	Fiscal Year 19	Fiscal Year 19	(Decrease) Increase Over 19
Norwood			
Cloth	33,506	41,311	(7,805)
Board	39,289	65,300	(26,011)
Endleafs	23,165	27,358	(4,193)
Cambric	9,674	9,461	213
Drills	1,646	(6,350)	7,996
Hollands	-	-	-
Thread	5,192	2,920	2,272
Supers	2,083	1,009	1,074
Kraft Paper Lining	5,912	3,396	2,516
Headbands	684	2,000	(1,316)
Adhesives	10,928	9,301	1,627
Wire	817	396	421
Stamping--Inks) Stamping--Foils)	2,277	5,020	(2,743)
Cartons	15,165	20,085	(4,920)
Kraft Paper	6,671	-	6,671
Tympan	995	691	304
Paper--Misc. Bindery			
Other			
Bindery Floor Inventory	13,348	19,073	(5,725)
Film, Blanket, Plate Inventry.	12,925	-	12,925
Mono-Lith Metals	57,673	57,673	-
Res. for Inv. Adj.	64,920	-	64,920
Misc. Sales Inventory	1,360	5,253	(3,893)
Book Inventory	405	216	189
Miscellaneous			
Rope			
Total Norwood	308,635	264,113	44,522

Exhibit 12-9

D. TOTAL INVENTORY ANALYSIS

In addition to the periodic checks on service levels, turnover and aging of items, a total inventory monitoring report is also needed. This report shows inventory dollars by item, by class of item and by total items. It should be compared with sales and cost of sales data to check for any seasonality effects. In this way problem trends can be caught quickly and corrective action taken. The report can be prepared weekly or monthly, depending on the dynamics of the business.

Even relatively unsophisticated inventory reports, such as the Raw Materials Comparison Report shown in Exhibit 12-8 and the Supplies Inventory Comparison Report shown in Exhibit 12-9, serve to identify trends and problem areas. The reports show inventory status by plant for a current month against the same month in the previous year. Using this simple system, the company (the largest manufacturer of elementary and high school textbooks in the country) reduced its raw material and supplies bill by $550,000.

E. PROFIT AND COST SAVINGS FROM REDUCED INVENTORY

The following discussion and financial schedules show how you can calculate the profit and cost savings resulting from inventory reductions.

Assume inventory reduced by $2 million—considering only 1.65 turns—going to 2.1 turns, where the industry average of close to 4.0 is reasonable.

The $2 million recovered increases profits and reduces costs due to the following items:

A. Out-of-Pocket Expenses—8–10% Value of Inventory
   1) floor space
   2) materials handling
   3) insurance
   4) obsolescence
   5) etc.

B. Interest—8%

C. Profit Opportunities—20–50%
   1) lost gross profit on out-of-stock fast movers
   2) additional gross profit on new products
   3) investment in cost reduction projects
   4) other investments, product lines, etc.

Assume conservatively A+B+C is 25%

COMPANY A—INCOME STATEMENT

Net sales	$22,884,666
Cost of goods sold	13,357,650
	9,527,016

Selling, general and administrative expenses	7,795,045
	1,731,971
Interest expense (19__—$168,829;	
(19__—$52,374) less other income	101,320
INCOME BEFORE TAXES	1,630,651
Taxes on income	709,752
NET INCOME	$ 920,899
Net income per share	$1.66

COMPANY A—REVISED

STATEMENT OF CONSOLIDATED INCOME

Net sales	$22,884,666
Cost of goods sold	13,357,650
	9,527,016
Selling, general and administrative expenses	6,795,045
	2,731,971
Interest expense (19__—$168,829;	
(19__—$52,374) less other income	101,320
Other income (cost saving and profit opport.)	500,000
	2,130,651
Taxes on income	926,768
NET INCOME	1,203,883
Net income per share	$2.17

COMPANY A

KEY PERFORMANCE MEASURES

Profit Rate =	$\frac{\text{Operating Income}}{\text{Net Sales}}$ =	$\frac{1.6}{22.9}$ =	7.0%
Capital Turnover =	$\frac{\text{Net Sales}}{\text{Net Assets (1)}}$ =	$\frac{22.9}{8.1}$ =	2.8
R.O.I. =	$\frac{\text{Profit Rate}}{\text{Capital Turnover}}$ =	$\frac{1.6}{8.1}$ =	19.8%
Working Capital Turnover =	$\frac{\text{Net Sales}}{\text{Working Capital (2)}}$ =	$\frac{22.9}{5.3}$ =	4.3
Fixed Asset Turnover =	$\frac{\text{Net Sales}}{\text{Fixed Assets}}$ =	$\frac{22.9}{.8}$ =	28.6
Inventory Turnover =	$\frac{\text{Cost of Goods Sold}}{\text{Inventory}}$ =	$\frac{13.4}{8.1}$ =	1.65

(1) Net Assets = Total Assets — Current Liabilities + Short-Term Debt.
(2) Working Capital = Current Assets — Current Liabilities.

COMPANY A—KEY PERFORMANCE INDICATORS

	Revised	*Present*
Profit Rate = $\frac{2.1}{22.9}$ =	9.2%	7.0%
Capital Turnover = $\frac{22.9}{8.1}$ =	2.8	2.8
ROI = $\frac{2.1}{8.1}$ =	25.9%	19.8%
Work Capital Turnover = $\frac{22.9}{3.3}$ =	6.9	4.3
Fixed Asset Turnover = $\frac{22.9}{.8}$ =	28.6	28.6
Inventory Turnover = $\frac{13.1}{6.1}$ =	2.15	1.65
NET INCOME PER SHARE	2.17	1.66

13

Avoiding Costly Pitfalls in Inventory Systems

A. THE CURRENT TRACK RECORD OF INVENTORY MANAGEMENT PERFORMANCE

The track records on successful inventory management system implementation are pretty dismal.

A national survey covering many firms in a wide variety of industries is presented in Exhibit 13-1. The data shown in the Exhibit are based on historical information covering inventory to sales ratios. This shows whether companies are improving performance by reducing inventories in relation to sales volume, or, conversely, losing ground by increasing inventories in proportion to sales. The tables in the Exhibit were developed from data on inventory to sales ratios published annually by several business publications. In the tables, a plus (+) mark indicates more efficient performance (more sales per dollar of inventories now than 10 years ago—an improvement in inventory performance). A minus (—) mark indicates the opposite (less sales per inventory dollar —a worsening of the inventory situation).

Several years ago, a major physical distribution consulting firm estimated that the potential in inventory reduction throughout U.S. businesses amounted to $33 billion. Since, as the table shows, significant improvement in some sectors are matched by sizeable losses in others, little progress is being made toward the $33 billion goal.

The computers available today, along with the advanced state of inventory management science and the availability of personnel capable of implementing inventory management systems, strongly suggest that the projected $33 billion inventory reduction is still a practical objective.

INVENTORY PERFORMANCE

MANUFACTURING & CONSTRUCTION; MEDIAN INDUSTRY + 6%	Better (+) or worse (-) inventory performance compared to 10 years ago (in percent)
Bolts, screws, nuts and nails	+ 51%
Metal stampings	+ 47
Fruits and vegetables, canners	+ 42
Cotton cloth mills	+ 36
Bakery products	+ 30
Machine shops	+ 26
Meat packing plants	+ 21
Lumber	+ 20
Bedsprings and mattresses	+ 18
Furniture	+ 17
Confectionery	+ 16
Footwear	+ 12
Drugs	+ 9
Farm machinery and equipment	+ 9
Coats and suits, men and boys	+ 6
Electric lighting & wiring equipment	+ 6
Paper mills	+ 6
Fabricated structural metal products	+ 3
Knitted outerwear	+ 2
Industrial chemicals	+ 1
Paints, varnishes, and lacquers	0
Soft drinks, bottled and canned	0
Airplane parts and accessories	- 2
Paperboard containers and boxes	- 2
Shirts, underwear, nightwear, men & boys	- 2
Hosiery	- 3
Motor vehicle parts and accessories	- 3
Hardware and tools	- 6
Work clothing	- 9
Breweries	- 16
Coats and suits, women's	- 17
Dresses, women, misses, and jr's.	- 21

WHOLESALING; MEDIAN INDUSTRY NO CHANGE	Better (+) or worse (-) inventory performance compared to 10 years ago (in percent)
Wines and liquors	+ 35%
Lumber and Building materials	+ 33
Electrical parts and supplies	+ 10
Hardware	+ 5
Automotive parts and accessories	+ 4
Drugs and drug sundries	+ 2
Meat and meat products	+ 1
Paper	0
Groceries	- 2
Electrical appliances, household	- 3
Paints and varnishes	- 3
Piece goods	- 6
Plumbing and heating equipment	- 7
Shoes	- 18
Tobacco and its products	- 31
Fresh fruits and vegetables	- 37

RETAILING; MEDIAN INDUSTRY - 5%	
Clothing and furnishings, Men and boys	+ 9%
Women's specialty shops	+ 3
Furniture stores	- 2
Department stores	- 5
Lumber and other building materials	- 5
Shoe stores	- 5
Grocery stores	- 6

Source: Drake Sheahan/Stewart Dougall, Distribution & Marketing Management Consultants, New York, N. Y.

Exhibit 13-1

The figures do show some small improvement in reducing inventories, but it is obviously occurring at a painfully slow pace.

In the course of presenting inventory management seminars across the country to hundreds of executives representing a wide cross-section of industries, two distinct attitudes on the part of participants have been detected. First there were those who had recently been given the responsibility to implement an inventory control system in their companies. Most of these men were eager to quickly plunge into the task. Then, there were those who had been in the inventory management game for some time. Some of the men from this group were merely trying to perfect their skills and confirm the fact that they were on the right track. Others had become disillusioned as original top management support and drive behind their efforts eventually waned and they found themselves unable to push through inventory reduction and control programs.

Part of the reason for the disillusionment reflected in the above informal survey is the naive approach taken by many firms in implementing inventory management systems. Because of lack of experience in this field, top management often places implementation responsibility at too low a level. The resulting lack of muscle when the going gets rough is in itself reason for system failure.

Characteristically, companies go through the following six steps as they progress through implementing inventory management systems:

1. Wild enthusiasm
2. Disillusionment
3. Total confusion
4. Search for the guilty
5. Punishment of the innocent
6. Promotion of non-participants

The above pessimistic picture is presented not to discourage those who wish to develop inventory management systems. It is designed to acquaint you with the reality of the situation. Very few computer salesmen or management consultants will do that for you.

Successful implementations are possible, even numerous. If you want to be among those favored few, it is necessary—no, vital—to appreciate the fact that inventory management systems are the most difficult to introduce of all management systems in use today. It is a management task which requires cooperation and communication across virtually all departmental lines. There are many "pretenders" in the system design area. Inventory system design requires a high degree of professionalism and acumen.

The following sections of this chapter are designed as a guide to success. The do's and don'ts which are presented are important individually and also when taken as a whole. What is suggested is, basically, a systems approach which by its nature leaves a smaller margin for error and failure.

Implementation programs are usually painful for some executives within the organization. Too often, the effort becomes the focal point of a contest between the "old guard" and the new "professional" managers. When these confrontations do occur, you may expect passive resistance at best and deliberate sabotage at worst. Under such conditions, chances for success are optimized only when one is aware of the pitfalls and when a careful plan of attack is prepared. So, if you want to be one of the winners, read on.

B. FACULTY SYSTEM DESIGN

1. Lack of Monitoring Systems to Measure System Performance

Several years ago one of the nation's largest distributors of automotive products implemented an inventory management system which was designed to reduce inventories and assure a high level of customer service (95%).

The system was introduced to regional distribution centers throughout the country at a cost of $250,000. The system's designers promised to deliver a 40% reduction in average inventory investment based on a 50% reduction in safety stock and a 20% reduction in working inventory. The system was based on ordering an economic order quantity whenever a pre-determined reorder point was reached.

Three years after the system was considered fully operational, a group of independent consultants were asked to conduct a survey of the system's results. It was found that the system had no performance measuring reports and the company's management did not know if inventory objectives had been achieved. The survey indicated that virtually no progress had been made toward the latter objectives. The system, it was discovered, was not being operated as designed. The major problem contributing to this lack of performance was that local distribution management developed their own version of the system's decision rules, and this was due to the following three factors:

(a) Lack of management discipline in enforcing the rules.

(b) Lack of understanding of existing rules and basic inventory management theory. Without some theoretical knowledge, the local inventory manager could not appreciate the effect his judgments had on inventory investment.

(c) The schedule for reviewing and reordering varied from one warehouse to another.

What was missing were customer service level, turnover and usage by value reports by distribution center, which, generated periodically, would have provided management with advance warning on performance failings. The consultant's study showed, for example, that in one major warehouse, in the sparkplug line, 20% of the items represented 94% of the annual usage (sales) and 43% of the inventory, while the other 80% of the items contributed only 6% of the sales but accounted for a whopping 57% of the inventory investment. Clearly, this situation reflected a very high inventory imbalance. Early awareness of this situation could have resulted in corrective action.

Monitoring customer service simultaneously with turnover is necessary to assure that a balance is being maintained between these sometimes conflicting objectives (i.e., maximizing service while trying to minimize inventories).

2. Fiscal vs. Calendar Month Reporting of Sales History

When designing the inventory management system, it is sometimes possible to smooth out what appears to be a high variability in monthly sales data. Recall from earlier discussion that the lower the variability in demand, the lower the safety stock may be.

Sales reporting by fiscal rather than by calendar month makes sales data appear to be more variable than it actually is. Consider the following example:

(A) UNIT SALES AS REPORTED BY FISCAL MONTH	(B) MONTH	(C) WEEKS IN FISCAL MONTH	(D) ACTUAL WEEKLY SALES (A) ÷ (C)	(E) CALENDAR MONTHLY SALES (D) × 4 WEEKS
280	Aug.	2.8	100	400
620	Sept.	6.2	100	400
400	Oct.	4.0	100	400
400	Nov.	4.0	100	400

Column E, Calendar Monthly Sales, shows a truer picture of what is happening to the marketplace and of the demands being made on inventories.

This illustration also highlights an important advantage in using computers for inventory management. Converting fiscal sales data to a calendar basis is a simple task for a computer even where thousands of items are involved. Manual conversion is too time consuming to be practical.

3. Establish Accurate Lead Times

Do not rely on purchasing, production, traffic or the Distribution department manager's subjective opinions on lead time. Lead time can vary significantly by supplier or by product. Common carrier service has deteriorated markedly in recent years, with inconsistent delivery times now the rule.

Rather than guess, take the time to random-sample all significant lead time considerations.

4. Provide for Consideration of Promotions

It is not unusual for special promotions to account for 10% of total annual sales.

An inventory management system should have the capability to evaluate the effect of additional, short-run demand due to promotions, on inventory requirements.

And, while we are on the subject, let's not forget to provide for communications between the Sales department and the inventory manager. It is first necessary to *know* that promotions are scheduled.

5. Forecast Monitoring

Some measure of forecast accuracy should be introduced into an inventory management system. This would be desirable whether the forecasts were based on statistical review and mathematical formulation or simply some individual's estimate of future sales.

Good forecasts cannot be produced consistently without the capability of monitoring them and making adjustments when necessary.

One firm, a giftware wholesaler who had relied on salesmen's forecasts for many years, was shocked to find upon objective analysis of the demand data, that these forecasts were, without exception, high. The result of these high forecasts was that the company was obtaining one turn a year on its inventory while it could have achieved four. Even as simple a method as monthly sales meetings to review forecast accuracy would have served to alert management to slow movers.

6. Exceptions Reporting

Another problem often associated with inventory management systems, particularly the sophisticated, computerized variety, is that the principle of exceptions reporting is not followed.

Where thousands of items are in the product line, it is difficult for the inventory manager to devote sufficient time to items requiring immediate attention if he has to give equal time to all of the items.

Exceptions reporting which shows just those items whose inventories, for example, have fallen below their reorder points or where additional production is required, calls attention to areas where action may have to be taken.

7. Use of Units in Calculating Usage and Turnover Rather than Dollars

A typical problem in inventory system design is that forecasts, particularly those generated by the Marketing department, are in dollars rather than in units. It is possible to get into trouble this way, particularly if a conversion from sales dollars to cost dollars is not made. Depending on what a company's mark-up is, use of sales dollars rather than cost dollars could result in ordering 20% or 30% more inventory than is actually needed.

To illustrate, if sales are forecast at $5,000 and selling price is $1 per unit, 5,000 units would be ordered. If unit cost were only $0.50, however, 2,500 units would be ordered for the same $5,000 sales forecast. Using sales dollars for the forecast, then, would have resulted in 2,500 units being left over in stock at the end of the period.

Experience also indicates that people tend to be a little more realistic when units rather than dollars are used for forecasts.

Using dollars has the added problem of inflation effects. When comparing usage or turnover from one year to the next, it is desirable to work in constant terms. Units provide that continuity. Dollars, on the other hand, are affected by inflation. Further, cost of sales may not rise at the same rate as sales price.

8. Order Data vs. Shipping Data Use as Forecast Calculations

In the interest of forecast accuracy, it is desirable to use demand data which reflects as closely as possible the real demand situation. Customer orders as received serve this purpose best. Shipments to customers do not include lost sales due to back-ordered quantities and, therefore, do not reflect the true demand picure.

A word of caution in using order data. Sometimes double counts are made on back-ordered quantities; e.g., where previously back-ordered items are sent through the system when stock is available as if they were newly ordered items. The system should be so designed as to avoid this counting of back-orders twice.

9. Output Report Frequency

Forecasts, order points and order quantities should be reviewed as frequently as is necessary to assure that the inventory system and its users are going in the right direction. Typically, new forecasts, related calculations and decision guidelines are updated and reviewed monthly. With faster moving items, such as those having a short shelf life, including all sorts of food products, a more frequent review might be necessary. The shortest review period would be one week.

Another area where more frequent review is called for is in items which are subject to a high degree of seasonality, such as sportswear or fertilizer.

10. Seasonality

Two types of problems may be seen in regard to seasonality. First, many firms do not introduce forecasting into their inventory management systems because they believe that their problems are too unique to lend themselves to solution. This is not so. Computerized systems can solve for the seasonality factor and provide reasonably accurate ordering strategies.

Second, some firms, feeling that they have insufficient data, start their systems by treating demand as being horizontal. Where seasonality exists, this approach results in having too much inventory on hand before demand peaks and too little on hand during the height of the "season." Where historical demand data are not readily available, it is strongly recommended that an effort be made to obtain such information. It may be necessary to key-punch old customer orders or to use a sampling procedure. Whatever effort is required, do it.

C. ORGANIZATIONAL CONSIDERATIONS

In an earlier chapter, much emphasis was placed on organizational structure and related functional responsibilities. These factors should be considered in the context of the following discussion.

Many inventory management systems do not reach their full potential because of a lack of attention to organizational and communication requirements. Of first importance is the fact that inventory management is not a part-time assignment, rather it is a full-time job. Too often, in fact, inventory management is no one's job. The whole subject covers many corporate functions and responsibilities; i.e., customer service, warehousing, purchasing, traffic, among others. Unfortunately, it is rare that one person is charged with pulling all of these functions together into one manageable entity. Instead, the typical situation finds the purchasing agent being told by the controller that he must cut down on some purchases one month. If the company is fortunate, the purchasing agent will check with someone in inventory control to find out what can be cut back without risking back-orders.

What is, too often, not recognized is that the goals of top management in terms of customer service, the goals of marketing in terms of customer service, the goals of finance in terms of turnover, the goals of warehouse management in terms of space, can all be brought together by selecting the proper ordering strategies, order points and order quantities. In this way, compromises between conflicting objectives may be achieved whereby everybody is satisfied.

It is important that one executive's desires not trip up everyone else, but, rather that individual desires fit into the total scheme of things. You cannot have, on the one hand, a salesman demanding that large inventories be accumulated for a promotion, with the result that only half that amount is sold, while on the other hand a financial executive demands 20 inventory turns per year causing intolerable stockouts.

There must be one person in charge of inventory management who is aware of all these important factors and how they infringe upon purchasing or production decisions.

D. KNOWING WHEN TO COMPUTERIZE AND WHEN NOT TO

The decision as to whether to computerize an inventory management system or utilize a manual approach, is largely based on economics. Firms having as many as 60,000 stockkeeping units do very well with manual systems. In such instances, the top 20% or so of the product line is given close attention while the remaining 80% is placed on a bin reserve basis. The latter system would typically require 14 people who would continually be updating inventory ledger cards and reviewing forecasts, order points and order quantities. You can buy a lot of computer time in exchange for that many people. Going to a computer in the above situation would bring the staff requirement down to two or three persons.

Even if a computerized approach to inventory management is economically feasible, it is strongly recommended that, as a first step, a manual system be applied. Before going to a computer application, at least a part of the product line should be handled manually. To do otherwise usually results in system breakdown and much disillusionment.

Many examples of uneconomic computer applications exist today. This has particularly been the case where smaller scale businesses have been acquired by large organizations. The new parent company sends in a team of experts who set up an "instant" system, generating numerous reports and costing as much as $5,000 per month to operate—a cost which most small firms could well do without. Then, the headquarters team leaves and the system is ignored—but the costs go on.

An example of an uneconomic computer application may be illustrated by a situation in which a beauty aids distributor found itself. Having annual sales of approximately \$3 million, this firm was spending \$2,500 per month on maintaining its own equipment. It generated stock status (not even inventory management) reports and a variety of sales and financial packages, most of which were worthless and not being used. In correcting this computer misapplication, the equipment was removed and a few special programs were prepared and run at a service bureau. The resulting monthly cost was \$400, and a more effective inventory management system was produced.

In another case, a giftware supplier was spending \$6,500 per month on an in-house computer system. It was found that the company was being charged for many reports which had become obsolete and were no longer being used by management. Merely eliminating these obsolete reports resulted in a \$3,500 reduction in monthly charges. After the remaining system was redesigned and an inventory management portion introduced, the total monthly bill was only \$1,200.

Now, assuming that you have a well-designed inventory management system, you know where you want to go and the economics are favorable, the following recommendations are made:

(For the sake of illustrating, assume that our hypothetical company has 5,000 items and that inventory management concepts are new to it.)

1. Establish a manually posted ledger card system showing order points and order quantities. Cover at least 20% to 30% in this way. This gives you an inventory management system which, if properly designed and implemented will enable you to very nearly attain the maximum inventory turns possible. Only a more sophisticated forecasting system will improve your turnover.
2. As a move to save you money, the next step would be to computerize customer orders and summarize these orders weekly or monthly. The respective item summary sales would then be manually translated into revised order points and order quantities and posted to the ledger cards.
3. Next establish computerized forecasting, even if this is only an average of the last three month's demand. The forecast results would subsequently be posted to the ledger cards in the form of revised order points and order quantities. This posting would be done on an exceptions basis, only where a change was indicated.
4. Finally, computerize all inventory status posting. Many companies mistakenly complete this step first. The error in doing that lies in the fact that this contributes nothing towards improving customer service or inventory turnover. The latter factors are improved with the introduction of better ordering rules. Set your priorities where the dollar savings and performance improvement lie.

Individual firms may not find it profitable to go through all of the four phases listed above. Before starting each phase, ask whether or not it pays to do so. Ask, "What is it costing me to do the job manually and how much will it cost me to utilize the computer further?"

Avoid relying upon hardware manufacturers in developing your inventory management system. Too often, their programs do not suit the user's requirements, rather the hardware manufacturer tends to try and develop your situation to suit his equipment.

Medium-size companies (up to $50 million sales) are beginning to use mini-computers to do customer order entry, general ledger, credit checking, inventory management and other work. This trend is taking place, and will do so at an increasing rate in the future, because of recent cost efficiencies in the micro-circuitry field. In-house computers are now possible which can be operated part-time, do not require enormous amounts of air conditioning and do not take up much floor space. Mini-computers have helped many firms reduce their total computer bill from service bureaus and in-house units by one-third. In the future, the mini-computer will become even more economical and thus permit smaller scale enterprises to take advantage of computerized inventory management systems.

Under current equipment technology, a general cost rule can be estimated. The most sophisticated inventory management system should cost no more than $400–$500 per month for the first thousand items, with small cost increments for each additional thousand items. As was shown in the above examples, it can cost a lot more if not done right.

E. ESTABLISHING DISCIPLINE IN INFORMATION PROCESSING

For many firms, the realization of the importance of accurate data input comes after system implementation, when the output reports start coming through. Only then do the distortions stand out in stark relief. Unfortunately, at that late stage, the task of researching the errors, arriving at solutions, redesigning subsystems and training personnel can be an enormous task. Even if the task can be accomplished, top management often becomes too disillusioned to continue the effort.

Why should data such as receipts, in-transit amounts, on-hand quantities and other statistics suddenly appear incorrect? The answer is largely that a great many companies process virtually meaningless data for years. Middle managers, aware of this, work around the problem and effect arbitrary adjustments to their records. Auditors who, for example, consider only the *total* actual (as taken in a physical count of inventory) inventory against the *total* book amount, help perpetuate sloppy record keeping. Inventory controllers and managers should be concerned more with the accuracy of each individual item going to make up the total than the total inventory figure itself. Recently, a Vice President of Distribution proudly presented an auditor's report which showed that out of a total inventory of 850 million units, there was only a one-half of 1% difference between the total physical count compared against the total shown on inventory ledger cards. Close examination showed variations between the record and physical count in virtually all items. Individual differences of ± 15% to 20% were common. The manager in charge of inventory control compensated for these individual variations by increasing safety stocks.

Another explanation of why data appears to go bad with the introduction of inventory management systems, is that these systems *use* the data whereas inventory control systems merely *record* statistics. Inventory control systems, for example, show perpetual inventory figures. An error might not be caught until a year-ending physical count is made. Inventory management systems frequently use the same perpetual record (the on-hand portion) to calculate weeks of supply. The weeks of supply statistics are then compared against order points. If, through receiving count errors, the on-hand record is understated, the resultant weeks of supply calculation will also be understated and premature reordering will occur. If, on the other hand, inventories are overstated, stockouts may occur before the shortage is recognized.

When, as is the case with inventory management systems, data is messaged to provide decision rules and guidelines, maintenance of the integrity of data is crucial. Electronic data processing managers have coined a succinct phrase which aptly describes the situation: "Garbage in, Garbage out."

F. SYSTEM DOCUMENTATION

Documentation is necessary whether one has a manual or a computerized system. It is important that continued use of the system not depend upon the availability of certain individuals. Good documentation should permit a new man to operate the system within two or three days.

There are two parts to documentation. One part outlines the technical details of the system. It shows all of the input and output flows. All numerical values of components of formulas are shown. Another, and possibly even more important part of documentation is "user documentation." This part shows how to apply the system to managing the company's inventories. It shows how people should react to what the output reports are showing (i.e., service level reports, turnover reports, purchase order budget report being exceeded). It shows people how to change things to move in the right direction, and generally provides users with a better understanding of their system.

14

Selecting a Warehouse Location Having the Best Profit Potential

Relocation from one site to another, or finding additional warehouse locations, is a necessity at one point or another in the career of many warehouse managers. This has particularly been the case in recent years where both industry and population centers have spread out from the traditional urban centers.

The warehouse manager may be solely responsible for recommending a site (new facility or leased), or he may be part of a management committee charged with developing a list of alternative locations.

We recognize that individual requirements may vary widely. One move may merely involve relocating from an obsolete, multi-story, leased warehouse to a larger, more modern location within the same city. In another instance, management may wish to re-evaluate the company's present multi-warehouse location arrangement and determine if their numbers and locations should be changed.

While specific cases require custom treatment, there are fundamental approaches and guidelines which can be used to simplify the warehouse manager's task. Often, merely being exposed to the way in which someone else handled a similar situation, a matter of a few minutes' reading, can make a month or more of data gathering and analysis a purposeful, rewarding experience.

The next few pages, then, are devoted to outlining two approaches to warehouse site selection.

The first is a manual procedure which will satisfy the great majority of situations which warehouse managers will be confronted with. The second is a description of the fundamentals involved in a computerized approach. Understandably, these procedures deal exclusively with the *economics* of locating a new warehouse. Usually, subjective, intangible factors also bear importantly on final site selection; i.e., management preference for commuting convenience; proximity to materials handling equipment servicing; quality of the local labor supply and so on. An extensive list of intangible factors is presented at the end of this chapter. It may well serve the warehouse manager as an item of supplementary review when making his presentation to management. His first concern and responsibility must be determining the economics of the situation.

A MANUAL ECONOMIC ANALYSIS

1. Determine Basic Requirements

a. First, a detailed study of storage space requirements should be made. If the present warehouse is located in an obsolete, multi-story, low ceiling building, you may find that you will require fewer square feet in a modern, high-cube facility.

   Don't think only in terms of present requirements. Request a five-year sales plan from management and relate this to your space requirements. Be alert to subtle changes taking place in your product line. Product proliferation—i.e., more colors and sizes of the same product—can, for example, result in greater inventories without a proportional increase in sales.

b. Determine floor loading requirements. One architect we know of specified structural floor strength at 150 pounds per square foot for the second floor of a new warehouse. He explained his arriving at that figure in the following manner, "That's what we used in a parking garage." Being unfamiliar with warehousing operations, he had no appreciation for the dense loads created with palletizing and high cube utilization. The actual requirement turned out to be 375 pounds per square foot.

c. Special labor requirements. Many warehousing operations also do fabricating and assembly work. If you need special skills and/or a large work force, inquire as to their availability at proposed locations.

d. Service level requirements. Management should be explicit in advising the warehouse manager of the maximum in-transit time which can be tolerated in shipments to key market areas.

   Some locations may have inferior common carrier coverage or be inconveniently located to transfer points.

e. What kind of supporting operations will be required at the new location? It is, for instance, less costly to lease already air conditioned space than to install the conditioning equipment yourself.

   Determination of factors such as the numbers and dimensions of truck docks; space required for offices and quality control laboratories; customer-will-call facilities; security requirements; etc., all bear importantly on site selection.

2. Select Logical Geographic Preferences

Most often, where a manual economic analysis is suitable, we find that the site choices are limited either by the size of the firm, the structure of its markets or management preference. As will be seen in the example detailed in the following paragraphs, the fewer the potential sites the more suitable is the manual approach. As our choices expand to cover numerous locations for a multiple warehouse complex, the more appropriate a computerized analysis becomes. We shall assume in this section of the chapter that the warehouse manager is in a position to select several logical locations. His task is to choose the single best economic alternative.

3. Total Annual Warehouse Cost Analysis by Location

The reader should bear in mind that our presentation follows the sequence which is recommended for his final report to management. Management is primarily interested in seeing the total cost profile first.

Listed in the left-hand column of Exhibit 14-1 are the major cost items to be considered in the analysis; viz, total annual freight in cost; total annual freight out cost; total annual leased warehouse cost; total hourly wage bill annualized.

It is important to note that we have not cluttered the analysis with costs which would remain the same from location to location. Such costs as inventory carrying costs, equipment maintenance expenditures, supervisory wages, etc. will not change appreciably from site to site, and would, therefore, not enhance our ability to select the best alternative.

The best economic choice shown in Exhibit 14-1 may be seen in a Bristol, Pa. location. That location has the lowest total cost profile ($68,996).

We shall now turn to a review of how the total cost figures are arrived at. Also shown are suggested methods of presenting the data for management review.

Exhibit 14-1

TOTAL ANNUAL WAREHOUSE COMPARATIVE COST ANALYSIS BY LOCATION*

	LOCATION		
Major Cost Item	*Roosevelt, L.I.*	*Wayne/Paterson, N.J.*	*Bristol, Pa.*
Freight In	$ 8,260.00	$ 7,180.00	$ 6,740.00
Freight Out	8,450.00	8,240.00	7,154.00
Warehouse Cost Based on 15 M sq. ft.	30,000.00	30,000.00	15,000.00
Hourly Wage Bill Annualized	37,336.00	44,720.00	40,102.00
Total	$84,046.00	$90,140.00	$68,996.00

* Includes only major cost items that would vary by location.

4. Comparative Wage Rate Analysis

The first step in arriving at the total annual wage bill, is to prepare a list of the number and respective job titles of all hourly paid employees.

Exhibit 14-2 illustrates how these findings may be shown. The most effective method we have found is to personally call firms in the same or similar industry, located in the geographic areas under consideration. This removes any question as to the reliability of the data.

Exhibit 14-2

COMPARATIVE WAGE RATE TABLE

	HOURLY WAGE RATES		
Job Classification	*Roosevelt, L.I.*	*Wayne/Paterson, N.J.*	*Bristol, Pa.*
Assembler	2.15	2.50	2.20
Assembler	2.00	2.50	2.20
Assembler	2.00	2.50	2.20
Checker	2.15	2.75	2.20
Shipping Clerk	2.75	3.00	2.39
Receiving Clerk	2.40	2.75	2.39
Helper	2.00	2.50	2.00
Floor Lady	2.50	3.00	3.70
Total Hourly Wages	17.95	21.50	19.28
Avg. Hourly Wage Rate	2.24	2.69	2.41
Total Annual (Hourly) Wage Bill	$37,336.00	$44,720.00	$40,102.00

5. Warehouse Space Cost Analysis

In the case of leased space, costs presented to management should include the per-square-foot rental charge and total taxes, since the latter will vary from location to location.

Commercial renting agents will be pleased to provide you with going rates in the various locations. Since the physical plants vary widely, it would be wise to visit several sites in each location. This will enable you to equate the quality of your facility to the rental cost. Of course, the per-square-foot costs finally selected should be annualized for purposes of presentation.

If you intend to build your own facility, the following costs should be obtained for comparative analysis:

a. Land cost.

b. Test borings if well water is necessary to support air conditioning system.

c. Soil consultant if required.
d. Site preparation costs.
e. Basic construction costs.
f. Basic engineering and architectural services.
g. Construction interest.

6. Common Carrier and Other Transportation Mode Freight Out Summary Analysis

Exhibit 14-3 presents a summary analysis which compares total annual common carrier freight charges by each of three locations. Exhibit 14-4 shows the detailed analysis used to generate the total dollar figures shown in Exhibit 14-3.

Exhibit 14-3

COMMON CARRIER

FREIGHT OUT SUMMARY ANALYSIS

I Bristol

(a)	Three-Week Total	=	$ 330.55	
(b)	Weekly Average	=	110.08	
(c)	$110.08 × 52 weeks	=	5,724.16	
(d)	$5,724.16 × 125% *			
	Total Annual Cost	=		$7,154.00

II New Jersey

(a)	Three-Week Total	=	$ 373.25	
(b)	Weekly Average	=	126.62	
(c)	$126.62 × 52 weeks	=	6,584.24	
(d)	$6,584.24 × 125% *			
	Total Annual Cost	=		8,240.00

III Long Island

(a)	Three-Week Total	=	$ 390.95	
(b)	Weekly Average	=	130.32	
(c)	$130.32 × 52 weeks	=	6,776.64	
(d)	$6,776.64 × 125% *			
	Total Annual Cost	=		8,450.00

* Factor to allow for Christmas peak shipments.

Exhibit 14-4

COMMON CARRIER

FREIGHT OUT DETAIL ANALYSIS

Period Covered: May 1, 19__ through May 21, 19__

All rates based on Class 70, Released at $0.50/lb.

		Shipped from Roosevelt, L.I.		*Shipped from New Jersey*		*Shipped from Bristol, Pa.*	
Order No.	*Destination*	*Applic. Frt. Rate*	*Total Frt. Charge*	*Applic. Frt. Rate*	*Total Frt. Charge*	*Applic. Frt. Rate*	*Total Frt. Charge*
21503–4	Chicago, Ill.	$10.05	$10.05	$10.05	$10.05	$9.95	$9.95
21501	Charleston, S.C.	9.63	9.63	9.26	9.26	9.06	9.06
21554	Indianapolis, Ind.	10.80	10.80	10.80	10.80	9.58	9.58
21605	Philadelphia, Pa.	3.70	12.24	3.48	11.76	2.95	11.00
21425	San Francisco, Cal.	12.95	12.95	12.95	12.95	12.95	12.95
21532	Natick, Mass.	10.80	10.80	10.65	10.65	8.15	8.15
21531	Newton, Mass.	10.80	10.80	10.65	10.65	8.25	8.25
21562	St. Louis, Mo.	5.33	56.12	4.69	49.57	4.28	45.10
21564	Ft. Worth, Tex.	6.86	68.60	6.53	65.30	5.78	57.80
21402–9	Washington, D.C.	3.97	39.75	3.34	33.40	2.64	26.40
21510–11–14–15	Indianapolis, Ind.	10.80	24.84	10.80	24.84	10.44	24.00
21589	Torrance, Cal.	12.95	46.05	12.95	46.05	10.97	39.00
21555	Oakwood, Cal.	12.95	38.15	12.95	38.15	10.97	31.30
21520	Chicago, Ill.	10.05	21.15	10.05	21.15	9.95	20.10
21630	Philadelphia, Pa.	8.15	8.15	8.15	8.15	7.65	7.65
21580	Orlando, Fla.	10.87	10.87	10.52	10.52	10.26	10.26
Total		———	$390.95	———	$373.25	———	$330.55

A typical three-week period is selected, during which all shipments from the present location (Roosevelt, L. I.) are recorded. Freight charges for two other proposed locations are then entered. At the end of the three-week period, total freight charges are shown for all three locations (Exhibit 14-4). Using the latter totals, we calculate a weekly average (Total Freight Charges ÷ three weeks) in Exhibit 14-3.

The average weekly dollar figures are then annualized for use in the "Total Annual Warehouse Comparative Cost Analysis by Location" analysis (see Exhibit 14-1).

In the above example, no cost schedule is shown for other transportation modes such as Parcel Post, REA and UPS. The zone rates for each of the above modes is the same for each of the three locations, and, therefore, need not be included in the final analysis. If the zone charges were different for any two of the locations, a similar analysis as outlined in Exhibits 14-3 and 14-4 would have to be effected. The results would then be shown in the Total Comparative Cost Analysis.

7. Common Carrier and Other Transportation Mode Freight In Summary Analysis

Exhibits 14-5 and 14-6 depict the methodology to be used in calculating the freight in costs. The discussion in Section 6, above, applies to the preparation and calculations involved here.

Exhibit 14-5

COMMON CARRIER

FREIGHT IN SUMMARY ANALYSIS

I Bristol

(a)	Three-Week Total	=	\$ 310.73	
(b)	Weekly Average	=	103.58	
(c)	\$103.58 × 52 weeks	=	5,386.16	
(d)	\$5,386.16 × 125% *			
	Total Annual Cost	=		\$6,740.00

II New Jersey

(a)	Three-Week Total	=	\$ 354.51	
(b)	Weekly Average	=	118.17	
(c)	\$118.17 × 52 weeks	=	6,144.84	
(d)	\$6,144.84 × 125% *			
	Total Annual Cost	=		\$7,180.00

III Long Island

(a)	Three-Week Total	=	\$ 381.24	
(b)	Weekly Average	=	127.08	
(c)	\$127.08 × 52 weeks	=	6,608.16	
(d)	\$6,608.16 × 125% *			
	Total Annual Cost	=		\$8,260.00

* Factor to allow for extra receipts during Christmas season.

Exhibit 14-6

COMMON CARRIER

FREIGHT IN DETAIL ANALYSIS

Period Covered: May 1, 19__ through May 21, 19__

		Shipped to Roosevelt, L.I.		*Shipped to New Jersey*		*Shipped to Bristol, Pa.*	
Receiving Report Number	*Shipped from*	*Applic. Frt. Rate*	*Total Frt. Charge*	*Applic. Frt. Rate*	*Total Frt. Charge*	*Applic. Frt. Rate*	*Total Frt. Charge*
498	Milwaukee, Wis.	$12.85	$12.85	$12.85	$12.85	$10.18	$10.18
499	Chicago, Ill.	3.85	326.40	3.58	303.51	3.31*	282.00
500	Trenton, N.J.	3.10	41.99	2.82	38.15	1.37	18.55
Total		———	$381.24	———	$354.51	———	$310.73

* Estimated Commodity Rate

8. *Analysis of Transportation Modes Used*

An important factor in site selection is proximity to the transportation modes to be used. Exhibit 14-7 shows a method for highlighting the company's relative dependency on each of the modes utilized. A typical month is selected and the number of shipments for each respective mode is entered. The right-hand column shows the percent of each mode's contribution to total shipments. (No. of shipments for a mode ÷ Total shipments = % of total.)

Exhibit 14-7

ANALYSIS OF TOTAL SHIPMENTS FOR APRIL, 19__

Type of Service	*No. of Shipments*	*% of Total*
United Parcel Service	248	47
Parcel Post	90	17
REA	15	3
Common Carrier	158	30
Air	19	3
Total Shipped	530	100

COMPUTERIZED WAREHOUSE LOCATION ANALYSIS

Computer warehouse location programs are designed to simulate cost structures. They measure the cost performance of each possible warehouse structure against the cost performance of existing warehouse structures. A comprehensive report is generated showing the detailed cost performances for all existing warehouses and for simulated warehouse combinations.

The cost performance measurements that are recomputed each time customers are reassigned to different warehouses are:

1. Cost of freight from warehouse to customers (reflecting adjustments for customer product mix, customer order size mix, contract hauling and drayage freight breaks).
2. Cost of freight on replenishments from plants to warehouses (reflecting adjustments for warehouse product mix and bulk replenishment ordering patterns). The bulk (carload, etc.) replenishment freight breaks reflect warehouse capacity-level changes.
3. Inventory carrying costs. Inventory is re-estimated using computations that consider annual replenishment requirement frequency, size of replenishment, replenishment lead times, updated customer service level requirements and product mix.
4. Warehouse operating costs (reflecting fixed and variable portions of cost at capacity level attained).
5. Gross profit on sales (reflecting warehouse product mix).

1. Problems Associated with Warehouse Location Simulation

Warehouse location simulation is a rather complex problem because of the large number of variables involved and factors to be considered. Some of the complexities encountered are:

a. Freight rate structures are complex and depend on many factors, such as:
   - —Classification of product being shipped.
   - —Size of order placed by customer.
   - —Size of replenishment orders and plant replenishment source.
   - —Method of shipment.
   - —Distance shipped.
   - —Amount of freight moved under a period contract.
   - —Amount of freight moved by non-regulated carriers in local areas, commonly referred to as "drayage."

b. Computation of distances between plant, warehouse and customer combinations is complex.

c. Cost structures associated with operating warehouses are stepped or semi-variable in nature.

d. Suitable warehouse location sites cannot be determined solely on the basis of cost factors.

e. Warehouses may have fixed maximum capacities, but additional warehouse facilities may be obtained in that same location.

f. All products are not available to all warehouses from the same source.

g. Warehouse performance can only be measured if it reflects the actual product mix being handled.

h. Safety stock requirements are dependent on the area being serviced.

i. A large amount of recomputation is necessary to test all possible warehouse structures.

j. Additional warehouses tend to stimulate additional sales, while warehouse reductions tend to diminish sales.

k. Sales patterns change over time.

2. Bases for Computations Employed in the Simulator

a. Freight rates:

Annual demands are classified by sales point into two distributions in percentage form:

1) 10 product classifications
2) 10 customer order sizes.

Annual demands are classified by warehouse into percentages representing 10 replenishment order sizes.

Rates pertaining to the actual method of shipment are used. Freight considerations are reflected in a "unit per mile" form, with percentage adjustments for other associated considerations. The amount of freight moved under a period contract associated with a given warehouse is restricted to the initial contract amount. It is not transferred to new warehouse structures. The amount of "drayage" associated with a given warehouse is not transferred to new warehouse locations when that warehouse is eliminated.

b. Distances between sales points and the existing warehouses servicing each sales point are given.

Distances between sales points and alternative warehouses are developed from a set of north and west coordinates associated with each warehouse and sales point being studied. The linear distances are computed and modified to approximate overland distances.

c. Operating costs are furnished for each warehouse, using a stepped, fixed-cost structure and a direct cost rate to approximate operating costs at various operating levels.

d. Actual test locations are provided by the program user. Typically, simulation output is not limited, and all alternatives are displayed. The cost performance measurements generated provide a convenient standard for ranking these alternatives, and the user can then choose the combinations that can best be implemented.

e. No allocation of sales points demand among warehouses is provided, but each warehouse is described with a cost structure that can handle several times its stated capacity. Capacity is expressed in terms of units of demand, thereby avoiding complex floor space or volume computations. Normal or existing warehouse capacity is given, but assignment of demand is assumed to be made to a warehouse location. It is assumed that an additional amount of warehouse capacity can be bought at that warehouse location. Unrealistic assignments of sales point activity to any given warehouse location are prevented by placing extremely high-stepped costs at the upper ranges of warehouse activity.

f. Product classifications used in the freight rate computations are identified by predominant plant source. Distances and lead times to different warehouse locations are given.

g. Product mixes for each warehouse are redeveloped after each reassignment of sales points to different warehouses so that:

   1) Product costs reflect plant sources.

   2) Product selling prices reflect actual sales point sales. It is assumed that for a short period after new warehouse structures are adopted, freight recovery policies cannot be changed regardless of warehouse used; i.e., sales prices actually established at sales points are assumed to be non-renegotiable.

   3) Working inventory estimates are based on average replenishment sizes pertaining to each product class. This same estimating approach is also used to evaluate the *existing* warehouse structure, even though average inventory balances may be available. This is done to provide a consistent basis for comparison of inventory estimates made for each simulated warehouse structure.

h. The program user has the option of computing additional safety stocks to reflect changes in areas covered. This is accomplished by computing an average lead time to customers, and adding it to replenishment lead time in the safety stock formulae.

i. Random number generation is used to select warehouse combinations for testing.

j. Any anticipated loss or gain in sales due to warehouse structure shifts would have to be estimated subjectively by the user in his final selection of the best alternative.

k. Changes in sales patterns can be tested. The user must make additional assumptions about sales points and the system must be resimulated using the new inputs.

3. Sample Printouts

Exhibits 14-8 and 14-9 show an initial system and one simulated system EDP printout, respectively.

The initial system covers an existing cost profit profile for a company having three warehouses. The total Net Profit equals 25.5%. Simulation No. 1 analyzed a proposed five warehouse complex yielding a Net Profit of 37.4%, a gain of 11.9%. Note that warehouses 4 and 5 are handling only 23% and 35% of their estimated capacity using this solution. This result highlights the effect that Freight-Out Costs have on the analyses.

It will be observed that a slight increase in Inventory Carrying Costs in Simulation No. 1 over the existing system was more than offset by reduced Freight Out Costs.

A final decision on what to do would involve consideration of other possible alternatives, the ability of the company to open and close facilities, the estimate of potential sales gains and losses by region and risk assumptions.

SITE SELECTION CHECKLIST *

Listed below are a number of factors which the warehouse manager may wish to review in his efforts to find a new site:

I. *TRANSPORTATION*

A. *Common Carriers*

1. Companies serving area and their geographic coverage.
2. Terminals and facilities.
3. Overnight service radius.
4. Schedules.
5. Rate structures.

B. *Other Motor Transportation*

1. Local rapid transit.
2. Nearest service to warehouse site.
3. Intercity bus service.
4. Taxi service.

C. *Commercial Air Service*

1. Airlines serving area.
2. Jet service if any.
3. Airport facilities—distance from city and warehouse site.

D. *Rail Transportation*

1. Lines serving city.
2. Classification territory.
3. Daily freight and passenger service.

* *Site Selection Handbook, Vol. 2,* copyright 1971, Conway Research, Inc., pp. 200–4.

WAREHOUSE LOCATION SIMULATOR

INITIAL SYSTEM EVALUATION — COST PERFORMANCE 11485910

WHSE NO.	FRT IN	FRT OUT	TOTAL FREIGHT	DOLLAR INVENTY	INVTY C.COST	ANNUAL $ SALES	PC CAP	COST OF SALES	GROSS PROFIT	G.P. PCT	WHSE OPER COSTS FIXED	WHSE OPER COSTS VARIABLE	TOTAL	NET PROFIT	N.P. PCT
1	1384518	6958329	8342847	310011	31001	11500000	115	8280000	3220000	28.0	10000	11500	52501	3162500	27.5
2	2236674	6296251	8532925	517481	51748	19500000	195	18699925	5580000	29.0	10000	19500	81248	4767520	24.8
3	4973709	14701163	19674872	404105	40411	14000000	140	13660000	4200000	30.0	10000	14000	64411	3555890	25.4
*	8594901	27955743	36550644	1231598	123160	45000000	0	40639925	13000000	28.9	30000	4500	198160	11485910	25.5

Exhibit 14-8

WAREHOUSE LOCATION SIMULATOR

SIMULATED SYSTEM EVALUATION NO.1

COST PERFORMANCE 16808000

WHSE NO.	FRT IN	FRT OUT	TOTAL FREIGHT	DOLLAR INVENTY	INVTY C.COST	ANNUAL $ SALES	PC CAP	COST OF SALES	GROSS PROFIT	G.P. PCT	WHSE OPER COSTS FIXED	WHSE OPER COSTS VARIABLE	TOTAL	NET PROFIT	N.P. PCT
1	1709076	11091603	12800679	347676	34768	13000000	130	9000000	4000000	30.8	10000	13000	57000	3943000	30.4
2	1376415	3000920	4377335	339591	33959	10000000	100	7000000	3000000	30.0	10000	10000	25000	2975000	29.8
3	3699403	8227101	11926504	367398	36740	10500000	105	6000000	4500000	42.9	10000	10500	40000	4460000	42.5
4	1043406	994911	2038317	114131	11413	4500000	23	2500000	2000000	44.5	20000	2250	30000	197000	43.8
5	1103739	1336623	2440362	177924	17792	7000000	35	3500000	3500000	50.0	20000	3500	40000	3460000	49.5
*	8932039	24651158	33583196	1346721	134672	45000000	0	280000000	17000000	37.8	70000	39250	192000	16808000	37.4

Exhibit 14-9

4. Shipping time to particular cities.
5. Carload rates.
6. Less than carload rates.
7. Piggy-back and other interchange services.
8. Demurrage.

E. *Highway Transportation*

1. Distance to nearest interchanges on Interstate Highway System—existing or proposed routes.
2. Access to other major highways.
3. Weight, height and length limitations.
4. Bond issues for new roads.
5. Toll roads and amounts of tolls.

F. *Trucking Service*

1. Companies serving area: local, intrastate, interstate.
2. Terminals and facilities.
3. Overnight service radius.
4. Schedules—trips per day.
5. Specialized equipment.

G. *Water Transportation*

1. Channel width and depth.
2. Depth alongside site, if on waterway.
3. Distance to channel, if not adjacent.
4. Lines serving area: schedules, rates, commodities handled.
5. Port facilities—warehousing, transit shed, storage areas, stevedoring.
6. Interchange facilities.
7. Seasonal limitations.
8. Insurance rates.

II. *LABOR*

A. *Labor Force Inventory*

1. Total labor force within feasible commuting distance.
2. Number and percent unemployed during last five years.
3. Seasonal variations in employment.
4. Skills notably abundant in area, or in short supply.
5. Management potential:
   (a) Your requirement for next five years.
   (b) Undergraduate and graduate programs of colleges.
   (c) Executive development programs in colleges for non-students.
   (d) Executive development programs of various associations and organizations.

B. *Wages and Hours*

1. Local wage rates by occupation.
2. Fringe benefits usual in area.
3. Average work week.
4. Local holidays.
5. Domination of wage structure by single company or industry.
6. Cost of living index.

C. *Productivity*

1. Other employers' experience as compared with their plants in other areas.
2. Absenteeism.
3. Turnover in various industries.
4. Accident rates.
5. Local practices on lunch periods, coffee breaks, rest periods.

D. *Unions*

1. History of unions in area.
2. History of organization attempts.
3. Time lost due to strikes in last five years.
4. External or local control of unions.
5. Calibre of union leadership.
6. Attitude toward automation and other technological change.

III. *BASIC MATERIALS AND SERVICES*

A. *Purchased Materials and Services*

1. Location of suppliers; delivery times.
2. Location of materials handling equipment repair facilities. Check firms' reputations for quality and timeliness of work.
3. Availability and cost of needed security services.

B. *Storage Facilities*

1. Public warehouses.
2. Bulk storage terminals.

IV. *LOCAL GOVERNMENT AND TAXES*

A. *Financial Condition*

1. Annual budget.
2. Sources of revenue.
3. Debt per capita.
4. Salaries of local officials.

B. *Civic Attitudes*

1. Attitudes of city officials towards industry.
2. Problem areas:

(a) New revenue sources.
(b) Public improvements.
(c) Redevelopment of private property.
(d) Urban plans to rehabilitate "ghetto" areas.
(e) Annexations.

C. *Civil Disorders*

1. History of riots.
2. Civic action to solve social problems.

D. *Local Taxes*

1. Property taxes—real and personal.
   (a) Tax rates for last five years.
   (b) Method of tax assessment and equalization.
   (c) Balance between tax loads on industrial, commercial and residential property.
   (d) Amount of tax-free property in area.
2. School taxes, if separate.
3. Local license taxes, if any.
4. Comparison of local tax load with services rendered.

V. *STATE GOVERNMENT AND TAXES*

A. *Financial Condition and Attitudes*

(Same as above.)

B. *State Regulations and Legislation*

1. Laws affecting incorporation of businesses.
2. Laws affecting out-of-town state corporations.

C. *State Taxes*

1. Personal income tax.
2. Corporate income tax.
3. Sales, use or payroll taxes.
4. Workmen's compensation tax.
5. Inventory, machinery and equipment taxes.
6. Vehicle and other license fees.

D. *Comparison of Total State and Local Tax Load with Services Rendered*

E. *Future Tax Prospects in the Area, in View of Needs for Major Capital Improvements*

The following is a list of sources of information for the above categories:

I. *Transportation*

Freight agents for airlines, railroads, truckers; bus and water transport lines; timetables; freight schedules.

II. *Labor*

Local employment agencies; state employment service; U. S. Department of Labor's Bureau of Labor Statistics; census data; other employers; local and regional development agencies; newspaper want ads.

III. *Basic Materials and Services—Chambers of Commerce, Industrial Directories, Local Employers*

INDEX